**COMPUTATIONAL TECHNIQUES
AS AN AID IN
PHYSICAL METALLURGY**

COMPUTATIONAL TECHNIQUES AS AN AID IN PHYSICAL METALLURGY

Proceedings of the conference on 'Computational techniques as an aid in physical metallurgy', organized by the Physical Metallurgy Committee of The Iron and Steel Institute, and held at Bodington Hall, University of Leeds, January 4-5, 1971

The Iron and Steel Institute

PRINTED BY Unwin Brothers Limited
THE GRESHAM PRESS OLD WOKING SURREY ENGLAND

Produced by 'Uneoprint'

A member of the Staples Printing Group   (UO701)

Contents

COMPUTATIONAL TECHNIQUES AS AN AID IN PHYSICAL METALLURGY: OPENING SURVEY

A. A. Greenfield

669.017:519.24

As a statistician I have a fair idea of the meaning of computation. But what is physical metallurgy? From colleagues I obtained the following definition:

'Physical metallurgy is the study of fundamental metallurgical processes such as deformation, solidification, transformation, and alloying'.

This contrasted with industrial metallurgy, which was defined as:

'The study of production processes such as extrusion, rolling, forging, and steelmaking'.

This distinction seemed to please the pundits. But it didn't please me! It surprised me that a profit-orientated industry would tolerate such fundamental research in its Corporate Laboratories, so I added a few words of my own to the definition: 'The effects on commercial characteristics', so that the definition became:

'Physical metallurgy is the study of the *effects on commercial characteristics* of fundamental metallurgical processes such as deformation, solidification, transformation, and alloying'.

Happily, I realized that this was what I had been aiding computationally for some years and I quickly found some examples to mention in my synopsis:

(i) the use of factorial experiments to estimate the effects of compositional variables on the strength and toughness of quenched and tempered steels

(ii) the effects of analysis, particularly additions of titanium and nitrogen, on grain-coarsening temperatures, impact energies, and yield strengths

(iii) the effects of residual elements, such as phosphorus, arsenic, and antimony, on toughness and strength with different tempering temperatures.

Then, some months later, the synopses of the papers to be presented at this meeting began to arrive on my desk and there seemed to be no question about my interpretation.

The author is with Bisra—the Corporate Laboratories of BSC

The paper by Messrs Johnson and Taylor on analysis of elevated temperature yield or proof stress data is clearly concerned with measurable commercial characteristics of steels. As a statistician I regret that the study did not include experimental design, although Mr Taylor may explain why this was not possible. It did, however, include three other major statistical ingredients: the construction of a mathematical model prior to data collection, the estimation of the parameters of that model, and the testing of those estimated parameters. Their analysis shows how computational techniques may be used to establish functional relationships between causal variables and the commercial properties.

It is a pity, therefore, that we are being denied the expected paper by Mr Hewitt, in which we should have been treated to a concoction of experimental techniques, multiple regression analysis, and linear programming to optimize commercial properties.

Other papers, according to the synopses, seem to be more concerned with physical metallurgy as defined without my adjustment. Since I do not have sufficient knowledge of that aspect of physical metallurgy, all I can do is to listen to those papers without making any prior comment—except for one: that it is my belief that whatever the subject being researched, whether it be physical metallurgy or cake-baking, the search for objective truth must always involve the testing of a hypothesis, and that most hypotheses can be expressed as mathematical models with estimable parameters.

And so with every paper I shall be looking for agreement with that belief.

My job at Bisra is to provide computational aids to a varied assortment of research scientists, including physical metallurgists. A classical research situation, encountered daily, may be described by the following procedure:

The objective is first declared as the optimization of a product or a process; the characteristics of that product are identified, precisions of measures of those characteristics are stated; and the control variables are identified, usually the compositional and process variables, with ranges and precisions. The objectives of the experiment are usually represented as a mathematical model relating the measures of the product characteristics with the control variables, and an experiment is designed so as to estimate the parameters of that model as precisely and as accurately as possible within the limitations of experimental costs. The objectives of an experiment must always be to answer a precisely stated question or set of questions. Almost always these questions can be stated in terms of a mathematical model whose parameters are to be estimated or perhaps compared with an alternative model. Sometimes the objective goes so far as to include optimization, but even this is a particular case of estimation.

Interest at Bisra in experimental design and analysis was fostered by the enthusiasm of Dr Duckworth when he was head of the Metallurgy Division. He had succeeded so far as passing on some understanding of two-level factorial and fractional factorial experimental designs, and some valuable technological advances came through research programs based on such designs. Each of the three examples I mentioned in my synopsis used a two-level fractional factorial.

This approach, you will agree, is a considerable improvement over haphazard experimentation, which is still all too common. The two-level factorial design does have its faults however, and one of my objectives has been to overcome these faults by devising a technique for producing the best experimental design for every particular case. This would not be possible without the computer.

Very soon it will be possible for the following dialogue to occur between our computer in Sheffield and a research metallurgist. The dialogue will be through a keyboard and typewriter. The user will first establish the date and research name, whereupon the program opens up a new data file under that name. It then begins to ask the user questions about his variables. Which are the dependent variables and which are the independent? The answers may be given as names or as numbers. Also identified will be the intermediate variables which, to the physical metallurgist say, may be worth recording to further his fundamental understanding of the subject, but from the black-box statistical aspect may be ignored. An example of this type of variable is grain size. It is not an independent variable from the viewpoint of experimental design because it cannot be controlled directly. Strictly it is a dependent variable because grain size is determined as the response to control or independent variables such as composition and process treatment. On the other hand it cannot be claimed to be a commercial characteristic of steel, although there are said to be relationships between the grain size and the commercial characteristics. So I call it an intermediate variable and seriously suggest that too much time and effort is wasted in thinking about it at all.

Having established the names of the variables and their types the computer will probe deeper. What are the anticipated ranges of the variable values? What interactions between independent variables and what curvatures might be expected for each of the dependent variables? What accuracies may be expected in meeting the specifications of independent variables? Are there any practical reasons for dividing the full set of observations into blocks? Which of the variables are quantitative, and of these which have continuous values and which have discrete values? Which of the variables are qualitative and how many levels of each quality are there?

The computer program will allocate a disk area to the research project and will store the information so far obtained. It will then produce the most efficient design corresponding to this information.

It is to this part of the program, the experimental design, that I have devoted some considerable research effort and experiments are now being designed in Bisra with the help of a design program.

The research worker would then be expected to follow the computer-printed design and return to the computer later with his results. The analysis programs will, however, take into account any missing, spoilt, or extra data. The computer will produce reports in the form of prints of the analysis, which it does at the moment, plots of contours, and sections of response surfaces. These will assist the user to determine whether to make further observations, in which case the computer would offer its advice on further observation points, or to produce a final report and clear the disk area ready for another user.

This may all sound somewhat whimsical, but it is on its way. In fact it already exists in the sense that if the statistician and the computer are regarded as a married unit, the research worker may approach that unit in exactly the way described.

Now some further remarks about experimental design and mathematical models. The objectives of an experiment can usually be stated in terms of a mathematical model whose parameters are to be estimated. The best experimental design is that set of combinations of values of the control or independent variables which will permit the estimation of those parameters with greatest precision, with least bias, and within allowable cost limitations. If you will accept that we now have the techniques for doing this, it is clear that we must start with the mathematical model.

In February, 1969, I attended another ISI meeting on the subject of 'Mathematical models in metallurgical process development'. Most of the contributions were preoccupied with the establishment of functions purporting to explain processes in terms of the underlying physics rather than with estimation of relationships which describe the general behaviour of the processes, and it became clear during the meeting that metallurgists and physicists often expected equations involving exponents, logarithms, roots, derivatives, and integrals because such equations always have the respectable appearance of physical laws. This general faith in apparent physical laws seemed to lead to a startling confidence in their extrapolation from the design of laboratory apparatus to the design of production plant.

However, even the least cautious speakers at that meeting admitted the need for checking their experimentally determined laws when applying them on different scales. And some of the more cautious even confessed that their mathematical models were not strictly true, but claimed that so long as the models were effective they were acceptable.

I therefore suggest that in respect of extrapolation, or any other form of prediction, there is nothing to choose between an unpretentious statistical relationship and an alleged physical law: neither is safe to use without checks. Another argument supporting this suggestion is that the establishment of these physical laws requires estimation of parameters by using observed values and, therefore, despite their esoteric mathematical formulations they must still be considered from a statistical viewpoint.

Every analytical function may be expanded as a Taylor series and hence approximated as a polynomial truncated at a convenient order. Less commonly realized is that polynomials lend themselves very favourably to evolutionary experimental design and analysis which leads quickly and economically to optimum values of the set of control variables. A striking revelation at that meeting was that most workers wasted a great deal of time and money studying models involving red-herring variables: there were, for example, two papers and a volume of discussion on continuous casting. These were devoted largely to the establishment of heat transfer models made extremely complicated by the presence of several phases, moving boundaries, and wall shrinkage. I submit that these are problems strictly for the fundamental physicist in university laboratories, not for industrial research workers. The claim that mathematical models built by such studies lead to a better understanding of the underlying physics and thereby justify extrapolation to production plant is not supported by the plea, repeated several times during the meeting, that 'even if the models are not strictly true they seem to work'.

It would be far more effective to concentrate on the variables with which the customer will be concerned, namely cost, operating ease, and other commercial properties. Perhaps there would be less academic glory from such studies, but there would be much quicker financial gain to our employers.

My true recommendation is not quite as extreme as that for I do not believe there is a dichotomy between the fundamental theoretical approach and the statistical experimental approach. There must be a mixture. But it seems to me that sometimes the physicist is blinded to reality by his theory. At a recent statistical course I was presenting the well established concept of randomization of observational order. One physicist present refused to accept this, insisting that in every experiment he had ever conducted he had had every variable under control; and he was able to measure every variable exactly.

I have no doubt, however, that the statistical design and analysis of evolutionary experiments offers considerable commercial advantages over the more esoteric techniques of mathematical model building and solution, which on the admission of at least one contributor to the Mathematical Models meeting, requires a significant admixture of serendipity and fiddle factors.

I am equally sure that the approach I recommend is preferable from the aspect of computerization since it is easier through this approach to define generalized computer programs that can be applied across a very broad field of research subjects. This is the approach adopted at Bisra in Sheffield, and it must be supported for yet another reason: that we do not have sufficient computer staff to particularize.

My outline of a computer program package for the automatic design and analysis of experiments following a dialogue between the computer and the user may seem far fetched to some people. But even further fetched is the idea of full and flexible laboratory automation. Yet this is now a short-term objective at Bisra.

Mr White will be describing the laboratory automation at Henry Wiggins where, in fact, he uses the same computer that we hope to have at Bisra in the next few months: an IBM 1800. This is a process-control computer for on-line application. I am looking forward to hearing Mr White's presentation because I am sure I shall benefit from his experience of the 1800. There is, however, an important distinction between the Wiggins application and the Bisra application. In the former, as I understand it, the computer is being used to control equipment whose process equations, or mathematical models, are fairly simple and well established. At Bisra we shall certainly have some applications of that nature, particularly related to routine testing work. Some such applications include the logging and control of creep- and tensile-testing machines, a mass spectrometer, an emission spectrometer, and a chromatograph. But every one of these cases involves the control of an instrument whose properties have already been established in a static environment. Much more challenging is the prospect of extending to the dynamic on-line situation the automatic design and analysis of experiments already described. In these cases we shall be logging data from and controlling processes whose properties may not be known in advance; the computer will establish mathematical models describing the processes and will improve these models as it acquires more data. Thus the computer will learn from experience, but rather more quickly than a human being can, although admittedly with some limitations.

The system will be entirely flexible so that even if we changed our operations overnight from metallurgy to hydroponics it would still be useful: within each laboratory there will be a wall-mounted plugboard and keyboard connected directly to the computer. The plug sockets will be labelled with types of signals that can be connected: analog input, analog output, digital input, digital output; and the permitted voltage ranges. Within the computer will be a suite of generalized data analysis, acquisition and control programs. The user in the laboratory will plug leads from his experimental process into the plugboard. He will have a conversation with the computer which is similar to that described for the off-line automatic design and analysis of experiments. He will signal the computer when the experiment is set up and ready to go. And it will go! The flexibility must be stressed: it will not matter to the system whether the experimental process under study is a miniature electroslag refining plant or a tomato plant, so long as the signal types, voltage ranges, and sampling frequencies are suitable to the computer.

You may think that this is building castles in the air. But it is not. If our plans go well we shall have this situation in a year. You may be concerned that the computer will smother the inventiveness of the research scientist. My belief is that instead it will stimulate it by accelerating the ability to test new ideas and by removing computational chores.

This is perhaps the most striking way in which computational techniques will soon be aiding the physical metallurgist. There are other important applications practised at Bisra. These include computational experiments such as process simulation, the study of continuous stress systems using the finite element method, and the study of crystallographic transformations by purely numerical techniques.

Possible future applications of computers to metallurgical research include the analysis and simulation of stochastic processes applied to the study of such topics as grain nucleation and growth, migration of grain boundaries and dislocations, crack propagation under fatigue, and the diffusion of hydrogen. Such applications would, however, need much higher mathematical ability than is usually found in the steel industry, or so I believe, if it is felt worthwhile to pursue such studies. Perhaps the later papers and the discussion will show that the expertise already exists.

DISCUSSION OF THE FIRST SESSION

In the chair: Dr W. E. Duckworth (Fulmer Research Institute)

The Chairman: If I may start the discussion by summarizing what Mr Greenfield has said, I would suggest that he has thrown out three challenges to physical metallurgists. The first is that in his view physical metallurgists can alway define the framework in which they are operating and thus can apply statistical and computational techniques. The second is that, left to themselves, physical metallurgists seem to want to explore the entire n-dimensional space of the parameters with which they are dealing, rather than try to find the shortest route to the object they are trying to achieve. His third challenge is to question whether physical metallurgists are prepared to subject themselves to the discipline of interrogation about these objectives by computer. The computer has no training in physical metallurgy and will not sympathize with your difficulties in blurred definitions. If the computer finds that some of your objectives are inconsistent, it points this out, and if some of your favourite experiments are irrelevant to your declared objectives, it points this out also. There are many funny stories about computers but they all basically come down to what the computer technologist calls GIGO—garbage in, garbage out. Are physical metallurgists prepared to make sure that they do not put garbage into the computer so that they do not get garbage out?

Dr T. Gladman (BSC, Special Steels Division): I am not claiming to be a pioneer of the intermediate variable, but I have certainly been associated with one of them over a considerable period of time, i.e. ferrite grain size. It is an intermediate variable. It is possible to do without it, but at the same time it does provide one of the only means of measuring another intermediate variable, namely dispersion strengthening. I would say that had we not used grain size measurements in niobium steels and vanadium steels, we might not have known of the existence of another important strengthening mechanism. The problem is that these two intermediate variables have different responses to the same independent variable, and it is only by measuring the intermediate variables that we can establish how a particular property (or dependent variable) will change.

You may say that it does not matter as long as there is some sort of relationship between the dependent and the independent variables, but it seems to me you could waste an awful lot of time if you did not know which of the intermediate variables was operating. I must therefore defend the intermediate variable. I see the intermediate variable not as an extra, but as a shortcut, in some cases, to the optimization of properties. In this way experiments can be based on understanding rather than empiricism.

Mr Greenfield: You are right, it may be, in some cases, but I am sure you will agree that people do waste too much effort on intermediate variables for their own sakes: I suspect that even in your case we could have established quite a satisfactory model representing your dependent variables in terms of the independent variables without considering either of your intermediate variables.

Dr Gladman: This is the problem. I regard the intermediate variables as the functional variables, i.e. the ones that control. If you miss these out, then the only body of experience you have is within the experiments you have done. If you have a fundamental—and this is coming back to the scientific side of things—model of the system that is expressed in functional variables, then you can extrapolate outside your own experience.

The Chairman: Specifications are not expressed in terms of these intermediate variables. There is no specification which says that steel should have a grain size between, say, ASTM 10 and ASTM 14, nor are there any which says there shall be a certain minimum particle size and a certain minimum particle spacing. Specifications deal in terms of production practice and composition. That is why Mr Greenfield thinks these are the variables which should be used when planning the experiment.

Dr Gladman: I would not even agree with that. The specification of the steel has little to do with the research that is involved in understanding, manufacturing, and treating the steel to produce the specified properties consistently. I do not think the specification of the steel *should* say anything about either the process variables or the composition of the steel. The commercial specification should be stated entirely in terms of the commercial properties required.

Dr J. Barford (CEGB): If I may paraphrase your contribution, sir, I think all you said is that my boss, however stupid he is, is a lot cleverer than a computer, and I do not think anybody would argue about that. Probably most metallurgists can accommodate a few more variables even than an IBM 1800, even if they cannot handle them quantitatively.

With respect to Mr Greenfield's admirable talk, I would say it had little to do with physical metallurgy.

It was concerned with the Bisra objective of optimizing certain physical properties from within a fairly closely defined set of boundary conditions; the pyramid in your analogy. This is much easier than the original definition of physical metallurgy, which Mr Greenfield dismissed, and which I believe is the true definition. It is a relatively trivial matter to optimize those variables in that sort of pyramid shape, because the pyramid is defined by limits of human knowledge in that field, but the limits of physical metallurgy are not defined by the human mind, they are defined by Nature, and this cannot be encompassed in the technique Mr Greenfield has been speaking about.

The Chairman: The danger in fitting new experimental data to an established theoretical model is that some of the data which does not really fit may be ignored as 'experimental error'. Serrated stress-strain curves which were subsequently found to be of great theoretical importance, were originally ignored on this basis. It can, therefore, be very revealing to let the computer have the first opportunity of constructing a theoretical model and we have examples at Fulmer where this proved extremely useful.

Mr Greenfield: Generally speaking, I think computational techniques *are* conceptually trivial. From the point of writing down a properly formulated model and thus providing some discipline to the research, to the experimental design, to the data gathering, and to the data analysis, the computer merely helps by speeding up the work and getting results out much more quickly. Later, by putting the computer on line, we should be able to speed up the process even more: we will be able to have much more data and much more rapid data collection, and this may lead to some form of lucky strike. I agree that the metallurgist, who is the expert, has to be there. You cannot leave it to the computer to detect the lucky strike.

Dr W. I. Mitchell (International Nickel Ltd): May I make a plea to get the semantics quite clear. Computational Techniques can mean two things: either the manipulation of numbers or the use of computers.

I am very grateful to Mr Greenfield for, two years ago, leading our laboratory into the paths of hillclimbing and getting to the top of the pyramid quickly, which produced very dramatic results, but not in physical metallurgy. Some of the important pointers in physical metallurgy have come more from the other end of computational techniques whereby simply connecting a stress-rupture machine to some form of data capture and further processing—and quite trivial processing—by the computer immediately showed up something which would not otherwise have been seen.

I think the automated laboratory can produce bigger and better pyramids for the physical metallurgist to look at, either with his nose to the commercial grindstone to get right to the flagpole at the top or, more helpfully perhaps, that some of us will still be left to find Tutankhamen's tomb by scratching around the bottom.

Dr J. Congleton (University of Newcastle-upon-Tyne): Following on from this point of semantics, should we not, right at the beginning, instead of using the term 'the computer' use the term 'the programmer' or 'the program', because there is a tendency to get the impression that one is dealing with the computer directly, but its power is all in the programming. If you have a dialogue, you do not have it with the computer, you have your dialogue with the programmer via and aided by the computer.

Mr Greenfield: It is not just a dialogue with the programmer, it is with the computer, with whom the person having the dialogue is linked. But in a sense this means he is having a dialogue with himself, just as, when a research worker comes to me and says, 'I want some help in designing an experiment', I say, 'Well, I do not know anything about the subject; tell me about it', and I ask questions and he tells me more and I go on asking questions, and after a couple of hours, I have not given him any information at all but he says, 'Thank you very much indeed, very helpful', so just by telling me the answers to questions which arise from his information, he has actually had a conversation with himself and sorted out some information. In this later case, he will feed it into the computer. So that it is not just a conversation with a programmer, although the programmer may stylize the conversation to some extent.

Dr Congleton: Is not this the most important thing? You can only be told that a set of experiments is going to be silly if the program is written so that the computer can do something to tell you that. You are so dependent upon what the programmer has asked the computer to do with your questions, or answers.

Mr Greenfield: You have to accept that the man who wrote the program did know something about experimental design.

Dr T. B. Vaughan (Imperial Metal Industries Ltd): I wonder if Mr Greenfield would care to elaborate just a little more on the computer which designs your experiments for him? The impression I get from listening to Mr Greenfield is that this will give you factorial experimental designs for the whole of the experiment in advance. Dr Mitchell mentioned sequential and evolution methods, both of which do find application, perhaps more on the industrial side than on the laboratory side, and I would like your comments on how your computer design technique will handle that sort of problem.

May I also raise the problems which nearly all industrial statisticians are concerned with, the completely unplanned problem of analysing plant data, or that ever-present person who comes along and says: 'I have a lot of data here. You might like to look at it and see if there is anything in it'. Perhaps you can comment on that.

Mr Greenfield: On the first point about the experimental design, the system will design experiments based on the type of variables you have got, whether qualitative or quantitative. The design depends also on the model which is provided, because the designer sets out to estimate the parameters of the model and these must be estimated with minimum cost. They must also be estimated with minimum variance and minimum bias.

Dr Vaughan: In fact, your experiments have not only got to specify all the variables but the model as well?

Mr Greenfield: In the generalized case, we would start off with a polynomial. Usually we find for initial experimentation, unless some information is known to the contrary, that a quadratic is a useful first approximation. Using that, we can put the volutionary aspects into the program so that after the design has been produced and data returned, the in initial model can be updated by re-estimating its parameters. Then, if the objective has been to optimize, minimize, or maximize, by using the updated model we can estimate the next best set of design points.

Dr Vaughan: So you can do a simplex type of experiment?

Mr Greenfield: Certainly.

Dr Vaughan: But there is no advantage in using a computer in this way, because once you set up your original simplex, then knocking out the lowest or maximum value and the computation of the next experimental point can be easily done manually in most cases.

Mr Greenfield: There is an advantage because you cannot use a simplex on qualitative variables at fixed levels, and you have to have some sort of factorial design, but you may be looking for interactions of various sorts which you believe may occur from a metallurgical point of view. You cannot have a simplex design with a set of quantitative variables which are discrete. You may feel that it is necessary at certain stages to estimate the quadratic effects. A simplex design is not adequate for this. You need a rotational design.

Dr Vaughan: And your system is to have all these alternatives together?

Mr Greenfield: Yes, it assists in producing these designs.

The Chairman: I suspect that there may be several people here who do not know the difference between a factorial, a simplex, and a rotational design. Can you very briefly explain?

Mr Greenfield: I shall first illustrate the factorial design. Suppose we have two variables, x_1 and x_2, which although continuous can be specified only at discrete values such as 1, 2, and 3 with no specifiable value inbetween, then we should probably form a factorial experiment.

This situation is represented in Fig. 1. It can be seen that observations can be made only at the points indicated by crosses.

The simplest form of a factorial design is the two-level factorial in which each variable is specified at only two levels: a high level and a low level. With two variables there would be four observations; with three variables there would be eight observations; and, generally, with N variables there would be 2^N observations. With a three-level factorial the number of of possible combinations would be 3^N.

If the variables were purely qualitative, such as test techniques or laboratories, you may again form the set of all possible combinations. Suppose, for example, there were three laboratories participating in a trial of two test techniques: there would be six possible observation points.

This sounds trite, but when we deal with cases in which there may be five test techniques, three methods of preparing a specimen, seven laboratories, and you have a technical reason for making ten observations at each observation point, the cost may be exorbitant.

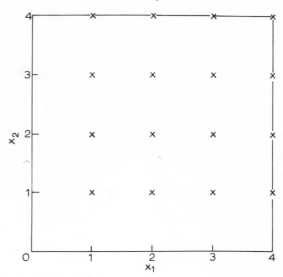

1 Factorial design

If you are just doing a carbon analysis and the cost is only 2p a time you may happily proceed with all observations. But if you are doing some other test which costs £10 000 a time, you may not want to do hundreds. You may wish to take a small subset of the total set of possible observations and the question is: which is the best subset to take?

Quite a lot of theory goes into answering this question and it depends, among other things, on the mathematical model of the system being studied. The physical metallurgist would not be expected to be an expert in this theory so he seeks help from the mathematician who may use the computer to produce the answer.

A simplex design may be used with fully continuous variables. It is an evolutionary technique that initially requires only $N + 1$ observations when there are N variables. Thus in the simplest case when there are only two variables, there would be only three observations initially. These are represented in Fig. 2 as the vertices of an equilateral triangle: A, B, C.

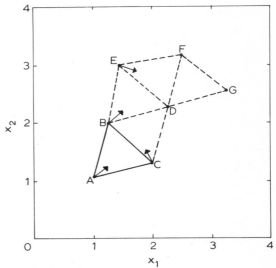

2 Simplex design

I shall use this two-dimensional, two-variables, diagram to illustrate the method although it can be extended to N dimensions.

Suppose you were trying to maximize the response to combinations of the two variables and having measured the response at A, B, and C you found the minimum response to be at A. You would then reasonably expect the line drawn from A perpendicular to BC to point towards the region of maximum response. The simplex method would be to make your next observation at D, located at the position of A reflected in BC.

Suppose that of the three points B, C, and D (A having been discarded), the response was minimum at C, the next observation point would be at E, which is C reflected through B and D. Similarly if B were then the point of minimum response out of B, D, and E, the next observation point would be at F.

This is a fairly crude form of experimental design but it is very rapid, especially if you have little computing aid.

Dr Vaughan: You are suggesting that, therefore, your computer will replace the statistician in the sense that the experimentalist will be instructed to perform a given experiment and return with the answers, which will then be analysed, further instructions issued, and so on?

Mr Greenfield: To some extent, yes, but the statisticians will be there for guidance. However, there are only three of us for a very large laboratory. This is my motivation. We cannot help everybody, but provided we get over the general method, and people continue to be cautious about the assumptions that are made in the design of the experiment and they continue to meet us for discussion and help, I do not see any danger. I think people will be sufficiently cautious about taking wholehearted advice from a statistician, but they can still discuss their problems.

Dr P. Duncumb (Tube Investments Ltd): There is one thing that I do not think anyone has actually said yet, although Dr Barford came close to it, and that is, what is the situation when one has a very small amount of data? How does one go about finding the relationship between, possibly, a number of variables, in a situation where the data you have is limited, or it takes too long to get as much as you would like, or as much as the statistician would like? I would suggest that, in that situation, one physical law is worth a great deal of data, and therefore it does pay, or can pay, to look very hard at the physics, or physical metallurgy, and try to establish some relations, even if somewhat tenuous ones, for the sake of saving oneself a great deal of effort during the experiment.

Mr Greenfield: It is true to some extent. It certainly helps later when you are analysing the data. The analysis package includes a facilities for transforming the data. We have quite a lot of cases where some knowledge of the physical laws has helped in giving transformations, in specifying transformations to the data before analysis, but we do get some funny people coming along. One research worker asked if I would look at his experimental design. He was going to try to estimate the effect of nickel on something, I do not know what, but he was only making one cast and I said 'Really, you ought to make more than one',

and he said, 'I thought the advantage of consulting a statistician was that he would tell me how to do it with less observations, not more'.

These are the extreme cases. We have some people who try to get far too much out of very small numbers of observations. Perhaps they have some faith in the physical laws and perhaps they are usefully applied, if they are well established, but they have to be well established. There are other people at the other extreme who blithely come along with thousands of observations and say 'What can you get out of that?'

Dr Duncumb: I see the dangers inherent in using physical laws which are not well established. I was worried by your statement: 'The system will be entirely flexible so that even if we changed our operations overnight from metallurgy to hydroponics, it would still be useful.' It suggests that you have a system which will just handle numbers, and therefore, if you change from metallurgy to hydroponics, you are not able to tell the system anything about the laws of hydroponics.

Mr Greenfield: The system is being designed so that the research worker, whether metallurgist or hydroponics engineer, can provide the computer with some information about relationships between variables to be used by the experimental design program.

Dr Duncumb: So you can tell its relations between the variables?

Mr Greenfield: Yes. One can type in relationships.

The Chairman: If a model is the best one available for the data, it would seem there is no harm in accepting it, but would it be tested?

Mr Greenfield: It would be tested on the observations. With some of the work we are planning now very carefully, we will be creating mathematical models as we go along. This is in a dynamic situation in steelmaking. The processes used in steelmaking have become so rapid now, they are very far from understood. A crude model is better than none from a computational viewpoint, and we will start off with crude models, but we will be gathering data and we will be updating models at every scan of the data points.

Professor J. Nutting (University of Leeds): The point I want to make is essentially the same as Dr Duncumb's, and that is, how far can you have your model divorced from physical reality?

Admittedly, I am making a statement that could be open to question because one does not necessarily know what physical reality is. This is part of the discussion you and I are having over certain other things, but one of the shrewdest comments I ever heard was that if it had not been for Andrade, we might well know something about creep.

When one starts thinking about that and the subsequent development to doing the thing you are suggesting, of developing polynomials, an example of this, of course, is the famous Nutting's equation, for which I have no responsibility whatever, for creep. That is a prime example of your approach to developing a polynomial which will give you some expression about the behaviour of materials under creep. Yet mathematically, it is an excellent model. It works and you seem to get something out of it. Physically, it tells us nothing, and from your analysis and from Nutting's equation, we would not know how to set about developing a new creep-resistant alloy, or a new creep-resistant material. It would not really give us any clue at all as to how to get to peaks of creep resistance, but on the basis of having a decent physical model of what was going to happen in creep, and then to start building up from this decent physical model, I think we have some hope of developing a better creep-resistant material. I am a little unhappy about this blind application of mathematical techniques without an awareness of the underlying physical principles.

Mr Greenfield: I do not agree. My approach to hill-climbing and optimization is that given a polynomial, or more simply at first a linear approximat, the coefficants are the gradients of the variable. For example, if you wish to find the best creep-resistant material, you would follow those gradients. The function would tell you, for example, what change to make in carbon. Thus coefficients do have physical meaning.

Professor Nutting: But the point is, if we take specifically that in relation to developing a creep resistant steel, not only are we involved with chemical composition as such, but we are involved in the widest possible range of heat treatments. Now, on your type of analysis, and with the two variables, this all looks very plausible, but when you get umpteen temperatures at which you can solution-treat the steel, umpteen different ways of quenching it, giving you various cooling rates, umpteen ways of tempering it, together with all you compositional variables, unless you have got some reasonable physical model on which to go, I would have thought that it would be more difficult to do your statistical analysis than to do the simple physical analysis.

Mr Greenfield: You have added only three more variables: quenching, cooling, and tempering. Geometrically the addition of three more variables would be three more dimensions. A simple linear model which may be admittedly crude is reasonably a first approximation of a situation. The gradients or the coefficients of those variables that you have estimated apply approximately over the range of your observations. In other words, if you have temperatures between 200° and 280°, your estimates will apply within limits which can also be estimated.

Mr Duncumb: May I give my fullest support to Mr Greenfield on that particular point.

Mr J. E. J. Bunce (UKAEA): I am a physicist who practises metallurgy, and I have experienced factorial experiments organized by a statistician without taking any regard for physical metallurgy which have ended in complete failure. I think one of the most important things in what you are saying, and a difficulty which you are going to have in this field, is for people to know in actual fact what are the relevant dependent and independent parameters to feed in.

One thing, which you mentioned in passing, which is likely to come into ever-increasing use in this field, is the use of regression analysis, because with prior data already accumulated, regression analysis can be used to great advantage to indicate to the researcher what are likely to be the important parameters to investigate.

The Chairman: This is, I think, the basis of a very long discussion and one of the marks of a very successful, stimulating speaker is one who provokes such a discussion that it must be drawn to a close. Equally, one of the marks of a successful chairman is one who detects when a thirst for knowledge is imperceptibly transformed into a thirst for alcohol. I think we have reached that stage and ask you to express our thanks to Mr Greenfield.

ANALYSIS OF LONG-TERM CREEP RUPTURE AND ELEVATED-TEMPERATURE YIELD OR PROOF STRESS DATA

R. F. Johnson and P. R. Taylor

The mathematical problems associated with the extrapolation of long-term creep rupture data and the derivation of realistic minimum properties for elevated-temperature tensile data have been treated using computerized techniques. A satisfactory method for analysing large quantities of mechanical test data has been developed. Data selection and various mathematical or statistical procedures can readily be applied and the results displayed in either graphs or tables. Operator decisions can guide the course of the analysis at all stages by the use of conversational programming techniques. The original system has been expanded to suit a variety of data handling problems in metallurgical and engineering fields and permits a more thorough examination to be made of the factors affecting property levels and their significance.

Mr Johnson is with the Swinden Laboratories of BSC, Mr Taylor is with Bisra—The Corporate Laboratories of BSC
620. 172. 251. 2:519. 28

In recent years considerable progress has been achieved in the international standardization of the rules for the construction of boilers and pressure vessels. Two of the principal design criteria employed are:

(i) minimum proof stress at elevated temperatures

(ii) the average long-term stress rupture properties at times varying from 100 000—250 000h.

In order to establish realistic levels for these properties for inclusion in the appropriate material standards, large quantities of data have been analysed. This has been done primarily on an international scale through the International Organization for Standardization (ISO) in order to develop international recommendations. Equivalent national specifications, although differing in certain detail, will in the main be aligned with the international standards. The UK holds the Secretariat of the Working Group responsible for establishing high-temperature properties, and the Data Analysis Group at Bisra has undertaken the actual analysis work involved.

Apart from the task of actually analysing the data, consideration has also been given to methods of analysis. In the case of proof stress data, this has involved finding a means of defining the minimum property in terms of the scatter band of data obtained, and in the case of the creep rupture data, of establishing the best means of estimating the required long-term properties from the shorter term data available.

The paper describes various aspects of the work undertaken at Bisra in conjunction with the UK and International Steering Committees. In the context of the present seminar, particular reference is made to

the use which has been made of the computer. With the exception of the handling of the complex polynomial expression for the extrapolation of creep rupture data and various curve-fitting techniques, the main use of the computer has been as a clerical aid in sorting and handling the information in a reasonable time. It is perhaps interesting to observe that one hour's computing time on a typical small computer system is £16, i.e. the cost of running one creep rupture test for 480h. One hour's computing time can extract from a prepared data file a considerable amount of information, perhaps involving a comprehensive 100 000h extrapolation of the data.

STATISTICAL NATURE OF THE DATA

One problem in dealing with the types of material property data under consideration concerns the statistical nature of the information. A variation in results obviously occurs due to macro- and even micro-variations in the composition of the material being tested, and the sensitivity of the material to heat treatment. In the case of long-term testing, variations can occur in the stability of the structure of the material over long periods of time. Added to this, variations occur due to the necessary wideness of commercial composition ranges and the tolerance which must be allowed in commercial heat treatment.

A further factor which affects the results is the actual condition of test and the manner in which the test is performed. In the last few years, however, testing techniques have been tightened up considerably and this has reduced to some extent the scatter associated with technique. In tensile testing, for example, one of the main factors established as being of importance is the rate of straining at the time the

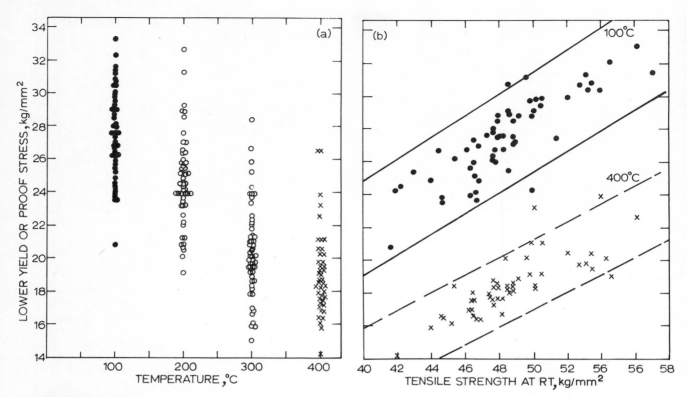

1 Dependence of elevated temperature properties on tensile strength at room temperature. Material: C-Mn steel

yield or proof stress is being determined.[1] This has now been internationally standardized. In the case of creep rupture testing, there are still certain features of technique which may affect the properties obtained. For example, it is difficult to ensure that the test conditions are maintained essentially identical for time periods up to eleven or more years. Also, economic considerations dictate that several strings of specimens should be tested in a single furnace, and the results obtained by this method may be different from the values obtained from a single specimen tested by high-sensitivity methods.

The data in question, therefore, form part of a distribution which at its simplest may be a normal distribution, but which is more probably represented by a complicated skew type of distribution. Figure 1a provides an example of the spread which may be obtained in proof stress values. The particular example quoted is for a carbon manganese steel supplied to one particular BS plate specification. Figure 2 illustrates the range of lives which may be obtained in stress rupture testing at a single temperature and stress level. When the data at other stress levels

are added at this temperature (Fig. 3), and considering that this situation exists at the other test temperatures, the magnitude of the problem can be appreciated. Figures 2 and 3 show the data presented in the classical manner, i.e. log stress v. log time; when the results are plotted on a linear scale (Fig. 4), the actual extent of the extrapolation necessary to reach 100 000h becomes apparent.

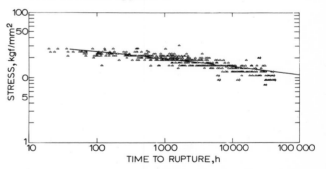

3 Creep rupture test results collected from many sources

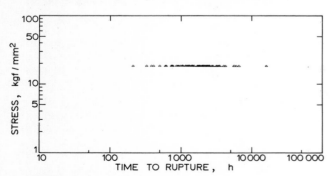

2 Distribution of test results at a single stress level

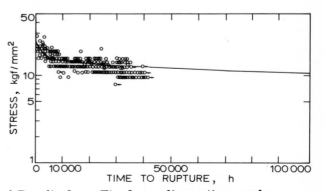

4 Results from Fig. 3 on a linear time scale

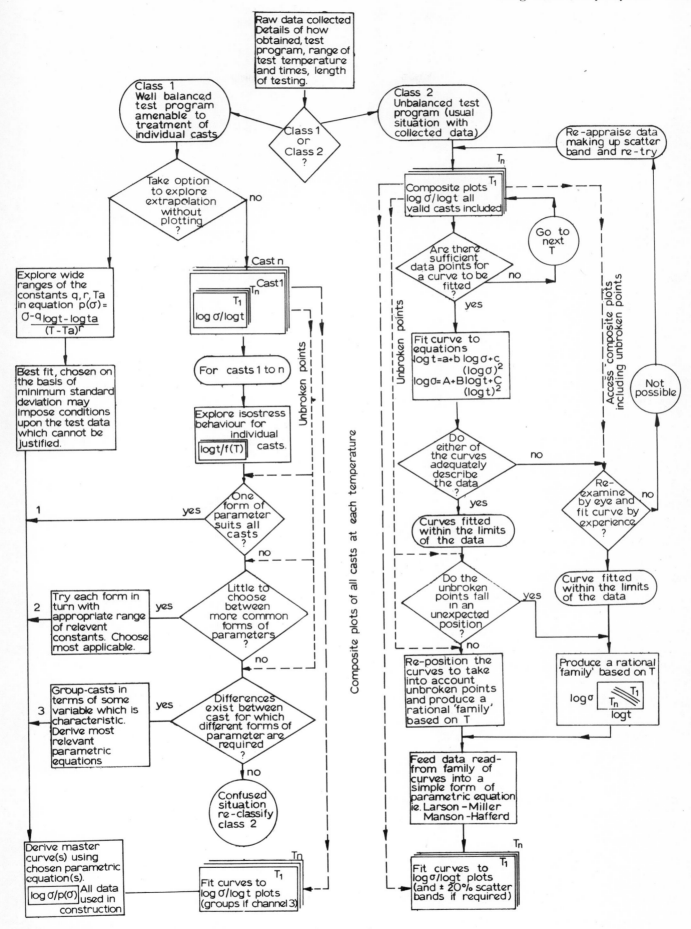

5 Flow diagram of the methods of treating creep rupture data

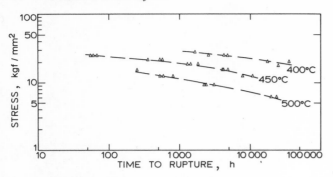

6a Mean curves fitted at individual test temperatures for iso-stress behaviour

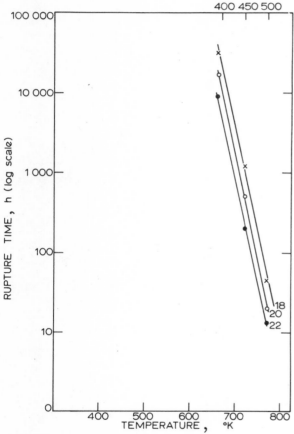

Form indicated   *MANSON HAFERD*
$q = 0$ $r = 1$ $Ta > T$ or $Ta < T$

6b Iso-stress behaviour obtained from 6a

ANALYSIS OF STRESS RUPTURE DATA

At present, the only satisfactory way of predicting long-term creep rupture properties is by testing a series of samples at several stress levels and a variety of test temperatures for times up to at least 10 000h (about one year), and preferably for longer. Time and temperature can be equated by the choice of a relevant parametric equation;[2,3] short-term tests at higher temperature may be used to predict longer-term behaviour at lower temperatures.

Extrapolation methods have been extensively used for producing long-term creep rupture estimates for two main reasons:

(i)   changes in design and the introduction of new or modified types of steel must be assessed ahead of being put into service, and a period of 11 to 25 years cannot elapse between development and use

(ii)   the cost of testing, which is about $3\frac{1}{2}$p/h, means that each test to 30 000h costs about £1 000.

The overall flow diagram for treating a set of creep rupture data is shown in Fig. 5. Along with the test results, composition, and heat-treatment details submitted for analysis there are usually notes on how the information has been obtained. These include the type of test program, the range of testing temperatures, the number of tests run, and the overall length of testing. This information guides the data analyst in the choice of approach for obtaining realistic long-term estimates.

Individual casts may be considered separately when they have been tested at a number of equally spaced temperatures with a similar number of specimens tested at each. The stresses applied should have been chosen to produce failures in from 100 to 10 000 hours or longer. Each cast tested in this way may be plotted individually in terms of log stress against log time for each testing temperature (Fig. 6a). Curves may now be fitted to each of these sets of results for the purposes of interpolation. From these fitted curves log time may be plotted as a function of temperature for given values of stress, (Fig. 6b). The shape of these iso-stress curves indicates the form of the parametric equation which suits the data being examined. Interpretation of the iso-stress behaviour

Isostress behaviour	Assigned values of constants	Resulting form of parameter
$\log t$ vs T (converging lines, steep)	$q = 0$ $r = 1$ $T_a < T_r$	$P = \dfrac{\log t - \log t_a}{T - T_a}$ (Manson-Haferd)
$\log t$ vs T (converging lines)	$q = 0$ $r = 1$ $T_a > T_r$	As above
$\log t$ vs T (crossing lines)	$q > 0$ $r = 1$ $T_a < T_r$	$P = \bar{\sigma}^q \dfrac{\log t - \log t_a}{T - T_a}$
$\log t$ vs T (curved lines)	$q = 0$ $r > +1$ $T_a < T_r$	$P = \dfrac{\log t - \log t_a}{(T - T_a)^r}$
$\log t$ vs T (curved lines)	$q = 0$ $r = -ve$ $T_a < T_r$	$P = (\log t - \log t_a)(T - T_a)^r$ (Manson-Brown)
$\log t$ vs T^{-1} (diverging lines)	$q = 0$ $r = -1$ $T_a = 0$	$P = T(\log t - \log t_a)$ (Larson-Miller)

7 Examples of specific forms of the generalized parameter equation of Manson

has been summarized in the paper by Harvey and May, [4] and is shown in Fig. 7.

If all the individual casts show a similar iso-stress behaviour then the characteristic form of the parameter obtained from Fig. 7 may be used in extrapolating the information. However, casts of nominally the same material can show differences in their iso-stress behaviour such that there may be little to choose between the more common forms of the parameter. The use of the computer system simplifies the dividing and grouping of the data and subsequent processing of the extrapolation so that if several casts show a similar iso-stress behaviour these may be treated as one group. The data may be grouped and regrouped easily until casts showing similar behaviour are identified together in terms of some common characteristic such as heat treatment conditions, or level of a particular element. The forms of the parametric equation relevant to the iso-stress behaviour may then be used.

Once it has been possible to allocate the values of the constants for the parametric equation it becomes a relatively simple matter to feed these into the computer program designed for handling this equation. The program used in the work of Bisra has made use of the relationship between absolute temperature and log time with the optional inclusion of a function of the applied stress, which general equation was developed by Mendelson *et al.*[5] This method equates time and temperature, and derives a general equation of the following form:

$$P(\sigma) \frac{(\sigma^{-q}\log t) - \mathrm{Log}\, ta}{(T - Ta)^r}$$

where $P(\sigma)$ is the parameter value, σ is the applied stress, t is the testing time, T is the test temperature in °K, and q, ta, Ta, and r are selected constants.

The program for handling the equation may be fed with actual data from the test program, and discrete values of the constants suggested after considering the iso-stress behaviour. Alternatively the computer program may be allowed to scan a chosen range of these constants and to pick a set of constants which produce a curve which is the best fit to the data. The degree of fit is optimized on the minimum standard deviation. The computer cannot be allowed to freely pick values of the constants with the sole object of finding the smallest possible standard deviation, since the chosen form of the equation could be a mathematically good fit while suggesting totally unrealistic metallurgical behaviour.

Unfortunately the greater part of the creep rupture test data which exists has been collected by individual laboratories operating their own test programs designed to answer immediate problems. Comprehensive test programs may not have been possible due to limited testing facilities. The wealth of data which is available from such sources both nationally and internationally is shown in Fig. 8. This represents an investment of about £3m.

Data collected from many sources often shows considerable imbalance between the number of tests performed at any given test temperature (Fig. 9). These data cannot be readily handled as individual casts, and are not amenable to being treated by the iso-stress approach. They are best treated collectively

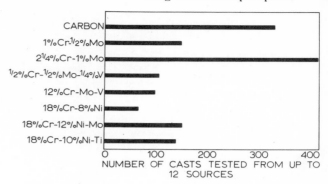

8 Some of the creep rupture data which has been collected

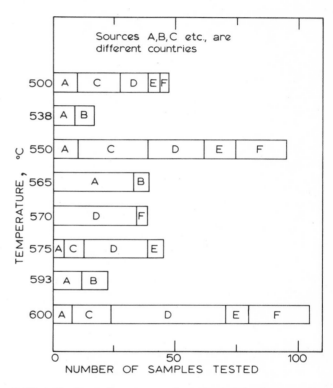

9 Distribution of creep rupture test data from various sources 2¼%Cr–1%Mo steel

by the production of composite plots of log stress *v.* log time, which includes all casts submitted plotted at each temperature as exemplified by Fig. 3. At this stage outlying test points can be seen and checked back to source for correction or exclusion, depending upon what explanation can be found.

The data analyst can now decide whether there are sufficient points for a curve to be fitted following the action outlined in Fig. 5. The curve which is fitted is in the form of a simple quadratic:

$$\log y = a + b \log x + c(\log x)^2$$

in which a, b, and c are constants chosen by the computer to minimize the sum of squares about the proposed line. This equation may be fitted with either time or stress as the independent variable.
In some cases neither of the forms of the equation fit, or adequately describe, the data. In these cases

the expertise of the analyst may be called upon to modify the curve slightly in the light of experience.

Tests still in progress and which have run for over 20 000h may be added to the plots (suitably identified as still on test), and used in visually weighting the position of the proposed mean curve. Their use must be based on an appreciation of the situation shown in Fig. 2. The presence of a large number of unbroken points may suggest a repositioning of the curves which have been produced without taking these points into account. When all temperatures have been considered and fitted with mean curves, the family of mean curves can be constructed. The spacing between individual temperatures in the family may be influenced by the amount of data at respective test temperatures. When the spacing is irregular (taking into account the distorting effect of the log/log scales) and this is not attributable to differences in the amount of data at the various temperatures then the analyst must use the knowledge he can gain from plots of selected individual casts which give an indication of iso-stress behaviour. In the treatment of such data, the values of stress, time and temperature from the family of curves produced as described above are fed into the program along with constants to give a simplified form of the parametric equation, i.e. Larson Miller or Manson Haferd. (Fig. 7)

All the work outlined above and shown in Fig. 5 may be performed manually, with perhaps the exception of the optimization of the constants in the parametric equation. However, the freedom to group and regroup the data along with editing facilities which are available in the computer system, greatly contributes to the speed of performing the analysis. The system is shown diagrammatically in Fig. 10. Using this system, a graph such as that shown in Fig. 3 may be plotted along with the mean curve, in under five minutes. This graph would take a skilled operator about five hours to produce and would not include the time for calculating the mean curve.

ESTIMATION OF MINIMUM ELEVATED-TEMPERATURE TENSILE PROPERTIES

Starting from a scatterband of data as shown in Fig. 1, the chief problem has been to find a satisfactory means of defining a minimum specification value which is realistic for design purposes. The fact that a factor of safety is applied to the value enables the problem to be approached in a statistical manner, and taking account of the risk in not meeting the properties which a steel producer can economically accept, it is reasonable to think in terms of a 95% confidence limit.

Various methods involving the concept of a 95% confidence limit have been used. These have been adequately described in the literature,[6,7] and it is not necessary to repeat the details here.

In developing the method which has recently been accepted as an international standard, the main factor was to recognize that the metallurgical variables which give rise to the scatter in elevated temperature proof stress properties have a proportional effect on the room-temperature tensile strength (on the room-temperature proof stress for austenitic steels). This is illustrated in Fig. 1b, from which it is seen that the relationship is linear over the range

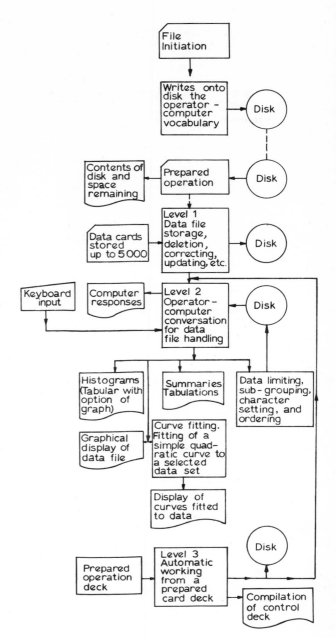

10 Schematic representation of the data analysis system

of tensile strength normally specified for a given steel type. The lines shown in Fig. 1b, indicate the 95% confidence limits, and the total spread in data at a particular tensile strength level, as indicated by the difference between the confidence limits, is only slightly in excess of 5 kg/mm² (49 N/mm²) at 100°C, and 4 kg/mm² (39 N/mm²) at 400°C.

The importance of recognizing this effect of tensile strength (proof stress for austenitic steels) cannot be over-emphasized. Earlier methods of defining the minimum proof stress value could give rise to completely unrealistic values being derived. As an example of this, the now out-of-date ISO Boiler Code simply considered a body of data in the form shown in Fig. 1a, without stipulating the number of tests required, and defined the minimum proof stress as the average value of the results available minus 2 standard deviations. Because some works produce material that on average gives tensile strengths biased towards the top end of the permitted range, there is clearly a danger that the Boiler Code method will give a high minimum value for the steel specification concerned. This is illustrated in Fig. 11, where obviously the realistic minimum is much nearer the lower left hand corner of the shaded area in Fig. 1*b* than has been actually derived.

The main requirement therefore, is that the definition of minimum proof stress should be related to the minimum specified room-temperature tensile strength, because a considerable percentage of any consignment could fall towards the bottom end of the specification range, for example, the composition may be towards the lean end; the tempering temperature may be towards the upper end; the section size may be heavy.

To allow for this situation, agreement has been reached that the specified minimum proof stress value at a given temperature be defined as the proof stress derived from the lower 95% confidence limit (of the proof stress *v*. tensile strength plot) at a tensile strength 3 kg/mm² (30 N/mm²) above the specified minimum tensile strength at room temperature. This is shown in Fig. 12. By definition, a small number of the data points involved will fall below the derived minimum level. In addition, the triangles bounded by the lines representing the lower 95% confidence limit, the minimum tensile strength, and the defined proof stress value (Fig. 12) will obviously contain data points. However, experience has shown that using the above definition, only 1% of the results actually fall below the defined value if the steel is supplied with tensile strengths corresponding to a normal histogram over the bottom 10 kg/mm² (≃ 10 N/mm²) over the specified tensile strength range. If the range supplied is wider, then this percentage will obviously decrease.

Using this method of presentation of the data, together with the definition of minimum value described, a clear picture is available of the reliability of the specified minimum. It is therefore possible to take this into account in deciding the safety factor which should be applied to the proof stress value for design purposes. In the event it has now been agreed that provided the lower 95% confidence limit is derived from test data on a minimum of 50 casts, then the previously used factor of 1·5 may be reduced to 1·4. In other words, it is recognized that some of the uncertainty attached to the minimum proof stress value has been removed.

It is not necessary here to give further details of the method or of its application in practice since these facts have already been published.[8] Also, as stated earlier, the procedure has been adopted as an ISO Recommendation and a parallel BS has already reached an advanced stage of preparation. It is interesting to note in passing that these standards also provide a means of verification, which when used in conjunction with the derivation procedure, eliminates the need for carrying out costly elevated temperature acceptance tests to prove the properties of a particular consignment.

The use of the computer in this work, as already mentioned, has been largely as an aid to sorting and handling the large quantities of data involved, and in carrying out the regression analysis. In some cases

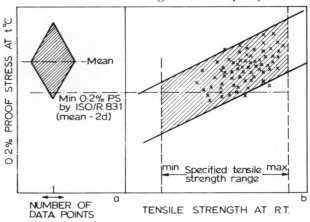

11 ISO/R 831 method-possible error due to biased selection of data

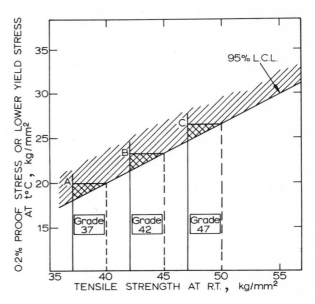

12 Defined minimum proof stress for any given minimum tensile strength

where the proof stress *v*. tensile strength scatterband appears abnormally wide, the computer facilitates the division of the data into different populations according to known metallurgical variables. Thus the scatter band and corresponding confidence limits can be readily obtained for any given set of parameters.

SUMMARY

The mathematical problems associated with the work described in this paper are readily solved using a computer. Extending the use of the computer to incorporate the sorting, editing, grouping, and display work involved in data analysis allows the analyst more time to concentrate upon the problem areas and the search for statistical techniques to extract useful information from the data files.

The system has been developed further to undertake assessments of corrosion and physical properties data. The significance of composition and other treatment variables has been examined and related to properties and behaviour.

REFERENCES

1   **R. F. Johnson and J. D. Murray:** 'High-temperature properties of steels'; 1967, London, The Iron and Steel Institute

2   **F. R. Larson and J. Miller:** *Trans. ASME,* **74**, 5, 765

3   **R. M. Goldhoff and G. Hahn:** 'Correlation and extrapolation of creep rupture data of several steels and superalloys using time-temperature parameters', ASM Tech. Rep. D8-3

4   **R. P. Harvey and M. J. May:** 'The application of time-temperature parameters for the prediction of long-term elevated temperature properties using computerized techniques', ASM. Publication No. D8-100

5   **A. Mendelson** *et al.*: 'Optimization of time-temperature parameters for creep and stress rupture, with application to data from the German Co-operative Long-Term Creep Program', NASA-TN-D-2975, August 1965

6   **B. H. Rose and H. G. Thurston:** 'Confidence in proof stress properties', Conference on Steels for the Power Generating and Chemical Industry, Tatranska Limnica, Czechoslovakia, Oct. 1967

7   **R. F. Johnson** *et al.*: 'High-temperature properties of steels'; 1967, London, The Iron and Steel Institute

8   **J. E. Roberts** *et al.*: 'Derivation of high-temperature proof-stress values for inclusion in steel standards', International conference on Pressure Vessel Technology, Delft, 1969, 987

USE OF A PROCESS CONTROL COMPUTER IN THE AUTOMATION OF MATERIALS TESTING AND ANALYTICAL LABORATORIES

T. K. White

This paper concerns the experience and activities of Henry Wiggin and Co. Ltd in the field of automation of their testing laboratories. The introduction and use of a time-shared process control computer in the automation of a large creep testing laboratory, tensile testing, and vacuum X-ray spectrometer control is described. Both the present real-time system, and possible future developments are discussed.

The author is with the Technical Department of Henry Wiggin and Co. Ltd

543. 438. 8+620. 172.
252. 1:62-52

Henry Wiggin and Co. Ltd operate 430 creep test units at Hereford as part of the comprehensive testing required to achieve the essential reliability of Wiggin alloys. The creep rupture test consists of loading a test specimen up to a specific stress at a set temperature and determining the time until the specimen fractures. In some cases it is also desirable to know the rate at which the specimen is extending or creeping under load; this is termed the creep strain test. About 30% of the machine-hours capacity is used for 'release' or 'production acceptance' testing which represents 75% of the tests carried out for durations of between 30 and 500h. A large amount of testing for research, development, and data purposes also takes place, for which tests can continue for 10 000–30 000 hours. Since no differentiation is made in the laboratory between 'release' and 'non-release' tests the problem resolves itself into one of 'production line' testing, since an average of 40 tests, 7 days a week are started.

Under the manual control system, the solution was to operate the laboratory on a continuous shift basis, i.e. four shifts of 3 men each to obtain 24h manning every day of the year. However, increasing production demands and more stringent testing specifications were stretching the capacity and capabilities of the testing laboratory. About four years ago it was felt that a major automation scheme was desirable in materials testing to increase capacity and achieve better technical control coupled with reduced costs and fewer operating staff.

After reviewing equipment available it was decided to evaluate the process control computer as being the most promising approach.

During the evaluation studies it was realized that the analytical facilities needed expansion and the purchase of an additional vacuum X-ray spectrometer made us consider automation in this field.

The complete automation of the Creep Laboratory, coupled with the control of two spectrometers presented a economically viable case for the introduction of a computer system. A detailed study showed that installation of the computer would lead to considerable reductions in both manpower and ultimately in costs, coupled with significant increases in productivity.

AUTOMATION BY MEANS OF A PROCESS CONTROL COMPUTER

The system is built around an IBM 1800 process control computer which was installed and commissioned during January 1968. This is a real-time computer system with a central processor unit (CPU) with 32K core storage and a 2μs cycle time. It has analog and digital inputs, digital output and various input/output devices such as a line printer, 'golfball' printers, card reach/punch, keyboard entry, and a graph plotter. The disk unit has three drives each containing 500 000 words of storage (Fig. 1).

The functions expected of the full system include temperature measurement and control, measurement of creep strain, calculation of tensile test data, vacuum X-ray spectrometer control, management functions, and an off-line computing facility. The computer utilizes the IBM 1800 Time Sharing Executive (TSX) Package to accomplish this.

The TSX organizes both IBM and our own written programs in precisely the best way to accommodate the calls of the system on a real-time basis, while utilizing any spare moments on any non-process work required. All program modules can be called in a random manner, either by other modules or by an interrupt servicing routine. This ensures that the most urgent needs of the plant are met immediately. Other programs which have been started are saved and worked off in order of their predetermined priority.

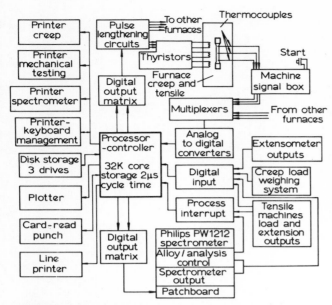

1 Process control computer system

In this system all input/output devices and process programs are assigned to one of 18 interrupt levels for execution, which is supervized by the System Director. An interrupt request is honoured if the level is not masked or if that level or any higher interrupt is already being serviced. Interrupt calls are queued if necessary, and when all have been serviced, the System Director brings in the non-process monitor which supervizes carrying out of non-process programs. Control passes back to the System Director when a process interrupt is raised. A combination of process interrupts and 'cycle-stealing' in the CPU allows considerable overlapping of input/output operations and computational work. Cycle-stealing is the term used to describe the high-speed transfer of data to and from a I/O device on a Data Channel. For example, if an input unit requires a core storage cycle to store data it has collected, the Data Channel with its 'cycle-stealing' capability makes it possible to delay the program during execution of an instruction and store the data word without changing the logical condition of the CPU. After the data is stored the CPU continues the program which was delayed by the 'cycle-stealing'.

Most of the process programs are written in 1800 Assembler, a low-level language similar in structure to machine language in which the binary instruction codes are replaced by mnemonic symbols and labels are used for other fields of an instruction. However, Fortran is widely used for non-process programs.

TEMPERATURE MEASUREMENT AND CONTROL

Automatic temperature measurement and control is provided for each of the three zones of all creep machine furnaces from the appropriate thermocouple attached to the testpiece. The temperature range covered is in general from 500°C to 1200°C and Platinum vs Platinum 13% Rhodium thermocouples are used exclusively. At the creep machines, plugs and jack sockets are used with low-mass prongs and inserts manufactured from alloys matching the thermocouples in use, to maintain accuracy, eliminate temperature gradients and spurious emf. The thermocouple outputs are then routed via a matching compensating cable to multiplexer units in the computer where they are switched through differential amplifiers which raise them to a level acceptable for conversion to digital form in an analog to digital converter. In order to achieve maximum performance from the system an effort has to be made to control the error present on the analog signal at the time of reading. Various sources contribute to the error in a given reading and these can be grouped into (a) basic System Error; (b) error due to noise; and (c) error due to common mode voltages. The first includes the effects of such things as the amplifier gain accuracy and stability, the amplifier zero offset error, the ADC comparator and reference voltage accuracy. Although the signal presented to the analog input system is in effect the true output of the thermocouple plus other signal components generally termed 'noise', it is convenient to separate the component caused by common mode voltages from this noise. Typical noise sources can be ac or dc from such things as 50Hz 'hum' from power lines, thermal emf, transient spikes due to motor commutators, switches, etc. or high-frequency interference. The majority of these problems can be overcome by good instrumentation practice such as physical and electrical isolation, avoiding areas of high electric field strength and shielding where possible. The effects of noise can be reduced by averaging several readings taken in a short period of time and this is in fact carried out in practice. Each of the 24 multiplexers has terminations for 60 differential inputs (20 machines), and is split between two analog to digital converters which run simultaneously, each sampling at a rate of 100 points per second. Under program control each multiplexer is sampled three times in succession and their average reading is used by the control program. The existence of common mode voltages could be a source of major error, and several methods are available for designing systems to minimize common mode errors. In this system the particular approach is to disconnect the common mode source for the period the particular point is being measured, termed the flying capacitor system. In this system the signals connected to the input connections charge up capacitors to the average value of the signal. Two pole, double throw mercury-wetted relays having a complete break before make action are used to disconnect the capacitor from the input and connect it to the amplifier—ADC for the few milliseconds required for measurement. The input impedance of the amplifier is high enough so as to make the capacitor discharge negligible during the measurement period. In order to provide open circuit detection a network is provided to give the 'flying capacitor' a discharge path should the input circuit become open. The control period is set at 32s, and at the beginning of each period both ADC's are set running. When each multiplexer block has completed its triple run the readings are processed by the computer to decide the amount of power required for the next period. The amount of power required by each furnace zone is computed using a proportional plus derivative plus integral control algorithm. Thus in simple terms the computer is acting as a time-shared three-term controller.

During the control period the scanning of the multiplexers and computation takes about 24s, but as the ADC's operate on data channels which do not need the computers attention once initiated, for most of the time the computer is free to carry out other jobs if required.

By the end of the control period a table has been set up giving the amount of future power output over the next period. The power output is based on the 16-bit computer word with all bits on for full power and no bits for zero power output. Although only 6% steps in power output are possible, provision is made to carry over any remainder from one control period to the next and the thermal coefficients of the furnaces provide a smoothing action. The computer output is in the form of 500µs pulses which are fed to a control cubide which has a matrix wiring arrangement for pulse routing. Each wiring matrix includes 16X and 16Y input terminals with cross connexions giving a possible 256 outputs per matrix. There is a total of six matrix arrangements, and 1 440 of the 1 536 available outputs are connected to pulse-lengthening circuit cards. These cards extend the computer pulses to one second duration and they are then fed via multicore cables to control units on each machine. The control unit is a triple unit with the thyristor driver, thyristor, and its associated equipment mounted together on one chassis. It is essential to ensure that the computer outputs do not interact, and to this end the input connexions are isolated from each other, and from earth at the pulse-lengthening card, and the one-second pulse is filtered before conversion by the driver unit to eliminate any possible crosstalk between channels. The driver unit uses the filtered signal to produce firing pulses to turn on the thyristor for the duration of the signal. The temperature control is between $\pm\frac{1}{2}$ and $\pm1°C$, and the automatic removal of temperature gradients allows the test to start within a short period of reaching temperature. This has considerable technical and economic advantages.

MEASUREMENT OF CREEP STRAIN

In some tests it is desirable to measure the extension, and thereby the strain, of a sample under test, and this is normally done by means of an extensometer. The limbs of the extensometer which are attached to the sample in the furnace extend outside where the relative movement is sensed by a pair of transducers. These are differential capacitive probes, which to obtain low temperature coefficients are manufactured from a controlled expansion alloy NILO alloy 36. The transducers operate into an automatic displacement indicator via a distribution system. The system is controlled by a computer but is returned to a manual mode when not automatically logging. The distribution system can accommodate a total of fifty junction boxes, each capable of driving 3 transducer pairs. When the process program calls for the displacement reading from a transducer pair the channel selector receives a signal from the computer. This signal causes the channel selector to scan sequentially until the required channel is recognized, whereupon it is decoded and fed to the junction box distribution system via 3 groups of wires. These are formed by a 10-5-3 wire system to select the correct junction box and transducer pair. When the pair of transducers have been selected, the automatic bridge balances and freezes. One second after re-

questing data the computer reads the displacement data which is provided in BCD form via the digital inputs. The automatic displacement indicator employs ac inductive divider bridge techniques, and the principle of null measurement to achieve absolute read out of displacement to an accuracy of 0·0002mm. 0·0002mm.

Logging of the displacement starts at the beginning of the test in a geometric progression from 2s up to 8s, continuing thereafter at 8h intervals until 300h, when a 24h interval is introduced. This continues until 3 000h, when the interval is extended to 96h, which continues until fracture.

At the termination of the test, the computer processes the data collected and in addition to printing out the strain-time information, can automatically plot the creep curve.

VACUUM X-RAY SPECTROMETER CONTROL

An analysis is made of practically every heat of each alloy melted in the company and for many elements it is convenient to use a vacuum X-ray spectrometer on account of its speed and accuracy of operation (Fig. 2). At present two Philips PW. 1212 vacuum X-ray spectrometers are in use, and like the creep machines have to be operated on a three shift basis. To analyse a specimen, the spectrometer directs a beam of X-rays onto its surface. This causes the emission of secondary X-rays of wavelengths characteristic of the elements present, the intensities of which, after separation by an analysing crystal, can be related to the amount of each element present.

For the analysis of unknown samples in a particular alloy, a series of standards of known compositions must be exposed first to prepare a set of calibration curves for the elements concerned. These were previously prepared manually and then the composition of the unknown samples read from them. This function is now carried out by the computer, which prints out the analysis of the subsequent unknown samples directly as a percentage of each element present, together with warnings if limits are exceeded or if agreement between duplicate runs is not achieved. This procedure is carried out each time the alloy is changed, and since Wiggin production is concerned

2 View of X-ray spectrometer

with over 60 different alloys of widely differing compositions, such changes are of necessity frequent. A facility is provided, termed 'save and restore', which allows an analysis to be interrupted without loss of data. This is especially valuable when a rapid analysis of a furnace melt before tapping is required.

The spectrometers can be operated under a wide range of conditions and optimum settings are required for each alloy. These important conditions are set by the computer upon recognition of the alloy code. Each of 15 analytical channels has the optimum setting of:

(i)   channel control

(ii)   analytical

(iii) type of counter

(iv)   X-ray anode voltage

(v)   X-ray anode current

(vi)   collimation

(vii) count control.

These are automatically made from data stored on disk. Some channels are outlets for more than one element, and a selector is provided for the desired element. Three different vacuum delay settings can be called and, a recycle facility for duplicate runs. The control console enables the operator to set the patchboard manually, leaving only the data collection facility, or even to work in a completely manual mode isolated from the computer. It is at the control console that the operator sets up the alloy code, and the number of samples on switches, and presses the 'complete' or individual element buttons. The 'complete' button selection instructs the computer to analyse on a preset number of elements. The operation of the 'standard' button gives a process interrupt which instructs the computer to read the information set up on the digit inputs and to program the spectrometer. On completion the ready light is switched on. The operator starts the spectrometer which runs through its analysis accumulating data from the standard sample for a duplicate run. When the exposure is completed and all data in, a calculation program is called which checks the standard

against remembered curves; the routine is repeated with the 'sample' button depressed, and unless the alloy is changed no further alteration to the control console is required. The 'autocalibrate' facility allows analytical runs to be made without the use of a standard, in which case previously stored information is used.

A serial method of data transmission is used whereby when a digit is set up an interrupt is raised by the spectrometer and the digit held until the computer signals the spectrometer to step on. This signal is given only after the digit has been read on the digital input. The format is fixed and a known sequence of digits is presented which is stored in the computer until completed where it is operated on by the spectrometer programs. Typical creep operator and spectrometer analysis outputs are shown in Figs. 3 and 4.

TENSILE TESTING

In order to read the load and elongation values during a tensile test, a tensile testing machine has been instrumented to provide this information in a digital form. Two digital shaft encoders are geared from the load pointer and the autographic recorder drum. The outputs of these two encoders are displayed on the front of the cabinet and are processed to provide an isolated BCD output for presentation to the computers digital input terminals. Readings are made every $\frac{1}{2}$s throughout the test, and at the end of the test the computer makes an automatic calculation of the various tensile parameters that are required.

MANAGEMENT FUNCTIONS

Automatic scheduling of tests is undertaken to obtain optimum use of the creep machines. Test data is inserted in the form of punched cards and comprises such information as the specification code, test temperature and stress, specimen dimensions (diameter or width × breadth measurement), identity, alloy code, discontinuation life if applicable, single or tandem test. A recent innovation to the scheduling programs stores the test information implied by the specifica-

3 Typical creep operator's output

4 Typical spectrometer analysis output

tion code in the computer and reduces considerably the amount of information necessary on the punched card. Some data is required for the analysis of testing carried out by the computer, at monthly periods, for costing purposes. When tests are scheduled all the necessary calculations are made by the computer and are passed on demand to the operators via the printers. To request a new test for a machine the operator presents a small negative voltage onto the bottom thermocouple input. This negative voltage is available on a plug which is normally parked under the thermocouple inputs. Upon sensing this condition during its next scanning period the computer extracts the next test for the particular machine in question and prints out such details as the identity and the load to be applied. Up to now the machine has been in what is termed the idle state and has been receiving a percentage of power necessary to keep the furnace warm: this state continues until the operator inserts the three thermocouples. This new condition is recognized by the computer which sends out a 'test has started' message, and the power output comes under the control algorithm. The integral term is not, however, brought into use until all three thermocouples are within $1\frac{1}{2}\%$ of the set point. When all three couples are within $1°C$ of set point and less than $1°C$ gradient a 'load machine' message is given and the operator puts the correct load on to the weight pan signalling to the computer that this instruction has been complied with. On receipt of the signal the computer responds with a 'machine loaded' message and starts timing the test. Automatic control of the test continues until fracture or discontinuation by the computer. In the latter case when the predetermined time has elapsed a 'test has been discontinued' message is given and the operator can then unload and strip the machine. When a fracture occurs the dropping lever or weight pan actuates a limit switch which shorts out a particular combination of thermocouples, a condition which the computer recognizes, signals 'test has ruptured' and reverts to the idle state. During the test the computer records the test temperature histories 'by exception', i.e. duration, extent, and time of excursions of temperature outside the set limits (usually $\pm2°C$) which are printed out at the termination of the test. Facilities are available for the operator to monitor any particular furnace and to obtain a temperature-test history to date.

NON-PROCESS UTILIZATION

The large quantities of tests involve even larger volumes of paperwork and an effort to reduce this was originally made by the introduction of an IBM 870 document writing system. Essentially this consists of a card read/punch unit fitted with a control panel which can be wired in a variety of ways to control the activities of an electric typewriter. In this way, decks or prepunched cards can be loaded on to the machine and the information on them typed into appropriate places or continuous preprinted stationery. Originally, it was intended to use the machine for paperwork within the laboratory only, but further evaluation studies suggested that optimum use of the machine would only result if the then administration system for cast-release testing was revised. Following negotiations with the other departments involved the current Wiggin cast-release system, was evolved.

Following installation of the computer, its further potential was realized and the production of the same paperwork was programmed as an off-line computing job. The present system has a great number of improvements apart from the speed of production, in that checks can be made on the sequence and correctness of the punched cards and conversions into other stress units are made automatically for tensile results.

Where possible the computer is used to assist management in day-to-day running of the laboratories. In particular the computer is used, in an off-line mode, to optimize and schedule the daily work load to the chemical, spectroscopic, and X-ray spectrometer laboratories. Considerable savings have been made in unnecessary transcriptions of data and a more efficient and economic way of working has evolved.

CONCLUSIONS: FUTURE DEVELOPMENTS

Automation is highly desirable in areas where there is a heavy commitment to bulk production testing particularly in these days of pressure to increase productivity and reduce costs. However, it is not only necessary to consider automating equipment, but to review and if necessary modify the management and operation of the equipment. This has been borne in mind throughout this particular project described here and its implementation has allowed a substantial reduction, both in manpower, and ultimately in operating costs, with a marked technical improvement over previous methods.

This system will be complete, with the integration of a hollow cathode spectrometer and another tensile machine, under the present computer configuration and TSX operating system. However if the multiprogramming executive operating system were adopted then it would be possible to extend the scope of this application and possibly broaden its use into new applications. This would be possible, as additional core and disk storage would be available as well as direct communication with the company-wide System/360 computer. In this latter application it is visualized that the testing laboratories could be linked more directly with the works processes which they are technically controlling.

In particular a feasibility study is being made of a fully automated production analytical laboratory built around a process control computer. A full cast analysis is usually derived from a number of different analytical instruments and a computer will be used to organize and control the instruments, and to manipulate the data generated. The possible laboratory could comprise

> X-ray spectrometers
> hollow cathode spectrometers
> carbon and sulphur analysers
> multichannel X-ray spectrometer
> multichannel optical emission spectrometer
> gas in metal analysers
> atomic absorption spectrophotometer.

At a central video display terminal data on a particular cast, when all analyses have been completed, would be edited and checked before release. A complete cast record would then be displayed or printed out at various locations within the plant. At a preliminary point a charge adjustment calculation may be made by the computer and conveyed to the Melting Department, when bath sampling is involved.

This development and other improvements in communications and reductions in process and holding times will increase productivity on plant and equipment, which are quite expensive capital investments.

While I do not necessarily advocate the use of computers for all automation, when several items of equipment capable of providing either analog or digital signals are producing masses of data for calculations of a routine nature then I think this approach should be seriously considered.

ACKNOWLEDGMENT

The author is grateful to the directors of Henry Wiggin and Co. Ltd, for permission to publish this paper.

DISCUSSION OF THE SECOND SESSION (part one)

In the chair: Mr J. Hewitt (BSC General Steels Division)

The Chairman: We have had two papers, the first of which presented the techniques of analysing high-temperature data, and the second which gave detailed information on the automatic collection and computerization of a particular testing setup. Dr Barford, would you like to comment on the collection of high-temperature data? This appears to be an aim in your particular industry.

Dr J. Barford (CEGB): Yes Mr Chairman, my organization is very interested in accurate knowledge of these long-term, time-dependent properties. Increasingly often we are thinking in terms of a design life of about 250 000h rather than 100 000h, and so we have a particular interest in the extrapolation techniques. One of Mr Taylor's slides showed an interesting, and perhaps important effect and I should be grateful for his comments. That was, on some of the graphs of log stress versus log rupture time there was a point of inflection at rather long times. This behaviour is presumably caused by some metallurgical effect occurring in the materials at long exposure times. Now if the effect is not manifest at test times up to that from which you make the extrapolation, but would subsequently occur, it seems that the extrapolated rupture stress might be pessimistic since it is influenced largely by results showing a marked decrease in rupture stress with time. I wonder if Mr Taylor has any evidence of this?

Mr Johnson: The effect described by Dr Barford may occur, and points to one of the weaknesses of the parametric approach. At the present time, however, there is no alternative to parameter techniques for estimating the stresses at the long times at which these are now required. Any errors involved in estimating these stresses obviously become less as the actual testing times increase, and for many of the standard steels, the data now extends to times of 50 000–80 000h; in a few cases even up to 100 000h. Consequently the 100 000h estimates, at the least, cannot be subject to much error. Metallurgical judgment must always be applied in interpreting values derived for longer times. Returning to Dr Barford's point, experience has shown that this phenomenon occurs fairly generally and is not confined to one particular type of steel. Dr Glen and myself published a paper at the 1967 Czechoslovakian creep conference which demonstrated the manner in which the 100 000h estimated stresses become progressively less with increase in test duration of the data. This trend applied equally to data examined from a number of different countries and was not affected by the particular form of parametric technique used for extrapolation. A similar effect has been shown in a recent analysis of international data, where, particularly in cases where the data now extends to very long times, estimates lower than those previously accepted have been obtained. The reasons for the effect are not clear at present. The fact that it does occur, however, emphasizes the need for the collection of long-term data.

Dr W. I. Mitchell (International Nickel Ltd): I would like to ask at this point whether anyone can make a guess as to whether the inflection, or the lack of ability to forecast here, is due to computational problems or whether there is a physical metallurgical change indicated by perhaps some more detailed look at the actual time extension curve.

Mr Johnson: There is no doubt that the inflections are due to metallurgical changes and these cannot be taken account of by present parametric techniques. The inflection may be peculiar to some particular time/temperature combination, which, however, may not be apparent under other conditions even though the time/temperature parameter is similar.

A Delegate: I would expect without any metallurgical knowledge whatever, to have a point of inflection, otherwise you would get a failure at some time with zero stress.

Dr Barford: The representation shown was of log stress $v.$ log time plots and therefore cannot fall off to zero stress. The difficulty is that representing this information on a log-log plot is for presentational convenience and may be visually misleading.

What worries me about the polynomial or parametric equations for representing metallurgical processes is that there are certain physical processes going on in the material over very long time periods which may significantly affect the properties; and, of course, none of these methods can *forecast* the occurrence or effects of these processes. For example, suppose you are testing a stabilized austenitic steel. For high-stress tests the material becomes warm-worked, giving a precipitate dispersion-strengthening effect. An alternative strengthening effect, currently being studied by Mr F. E. Asbury at CERL, arises from the initial supersaturation of carbide-forming elements; but this latter effect is mainly operative at low stresses insufficient to cause warm work. The warm-work effect persists for longer times than the super-saturation effect, but it is difficult to separate them quantitatively. Clearly, the presence of such complex metallurgical effects at long test times greatly aggravates the difficulties in extrapolation.

Mr Johnson: A further difficulty is the fact that the metallurgical behaviour of a given material can vary from cast to cast. This means that the shape of the rupture curve for individual casts of nominally the same material may vary, some showing a point of inflection, others not. In establishing the long-term properties for the purposes of design, it is obviously necessary to take a reasonably broad view of these effects, otherwise our task would be impossible.

Dr Barford: It is all very well for a materials producer to tell plant operators to take a broader view, but it is no great consolation if he is also told the stress to rupture has to be downrated from early estimates of the rupture life. The design factor applied to the 100 000 rupture stress is commonly 1·6. This has to take a lot of things into account other than material scatter. It has to take into account all the unknown stress involved: welding stresses, fabrication stresses, thermal stresses; and when you have taken all these factors into account as well as the scatter in the material properties, you might be working on a very thin line, and while the operator is always very glad to know what the most accurate extrapolation is, it is a matter of some concern if these have to be downrated.

Mr Johnson: We now feel fairly confident about the levels of properties which have recently been established. This confidence has been recognised in the draft ISO Pressure Vessel Code where the safety factor on the long-term properties has been reduced from 1·5 to 1·3.

Dr Duckworth: I do not think the speakers are too clear whether the drop in minimum values is due to the actual curve being different from the extrapolation or the scatter being different from the predicted properties, or whether it is a bit of both.

Mr Johnson: We are talking about average properties rather than minimum properties, so therefore the scatter in the data has little or no effect.

Mr B. Plastow (CEGB): I think it is probably valid to accept that the parameter plot represents the best prediction one can get from the test data which is included in deriving it. I would like to ask Mr Taylor: if he divides his test data into time intervals, does he get the same equation for each interval? For example if he looks at the data between 10 000 and 1 000h does he get the same result for that interval as for the whole data? If you take this a stage further and look at the data from 100 to 1 000h and predict the 10 000h data, do you get the same parametric equation when you use data exceeding 10 000h to predict the 100 000h life?

Mr Taylor: The simple answer would be no; but it would require qualifying, to say what do you mean by 'the same'? To within what sort of tolerance would you say the answers were the same? The problem is that the actual optimization of the parametric equation leaves some freedom or loophole for differences to arise. I am thinking of an example we did quite recently, where we did deliberately ignore long-term tests. We took a set of data which had been well tested up to 40 000 or 50 000h and deliberately truncated it at 10 000h. We used just that portion of the information and predicted the average stress for rupture in 100 000h. We then took the whole data set and followed the same steps in the analysis and got that prediction. The two values were only slightly different, i.e. about $\frac{1}{2}$kgf/mm^2 different (1 t/in^2) which I would consider to be close. Whether this would be acceptable or not depends on what percentage of the value at 100 000h this difference is. This depends on whether you are looking at the high- or low-temperature end of the testing range.

Mr Plastow: This is what I am trying to derive, what scatter do you predict at 100 000h? It obviously depends on the amount of data you are feeding in.

The Chairman: Would we not generally accept that the longer the time interval you looked at it, the more accurate you would get?

Mr Plastow: What I would like an opinion on is: should we neglect all the results below say 10 000 or 5 000h? Could these results be misleading?

The Chairman: Specifically, is the scatter in the initial period to 1 000h such that it is best to ignore that information?

Mr Johnson: This point has been raised in relation to the analysis of international data, and the decision taken to ignore all test results below 5 000h.

Mr Taylor: We would ignore casts which had only been tested to a maximum of 1 000h. If a particular testing laboratory examining a cast of steel failed to test beyond 1 000h at any temperature, we would ignore their work on that particular cast. However, in the presence of longer-term tests we would not take out of the picture any of the short-term results. We would not take out all the points at less than 5 000h simply because they were short term and say that we would not include them in any of our parametric assessments, unless we found that they were introducing a lot of variation which was not substantiated by the longer-term tests. We have tried to make the approach objective by getting the computer involved in this work and by trying to define and simplify the decisions that are necessary. It is still a highly subjective situation, where the expertise of the individual looking at the final picture greatly influences the final results. The expertise involves looking back to see how the results are being influenced by certain bodies of points, the overall spread of the points and by the lack of data at certain temperatures. If some temperature is out with the rest or if there are only a few test points or there is something obviously peculiar, the analyst would go back and check this area thoroughly. It is not a simple process where the 'analysis machine' is switched on and the answer comes out. There is quite a lot of deliberation involved and problems are considered as they arise. Some answers have been found but there are still big question marks hanging over certain areas.

Dr P. Duncumb (Tube Investments Ltd): Would Mr White comment on the amount of effort he has put into the software side of his installation, the programming? It is relatively easy to get the hardware costs, but how much has gone into the programming?

Mr White: We estimate that the software that went into the project was about ten man years. This was provided by our Computer Services Department, which is in fact a company service, so fortunately we did not have to include the cost in this system.

Dr Duncumb: Over what period was that spread?

Mr White: This was over a two-year period. There was the project team which varied in number. There were about four people working on the project. Two of these were technical people who were trained in programming techniques and two were at that time data processing programmers who had to be retrained in the process control work and in technical matters. Obviously, there was support and backing from IBM on this and this is one of the reasons why the 1800 was chosen. At that time, they had the only really efficient operating system available and this is what makes the system an efficient operating system, which controls the whole computer system.

The Chairman: Could I ask Mr White what sort of advantages he sees, having used this setup for over 12 months now? I am thinking that one of the major points which always arises in questions about data and indeed normal ambient temperature test data is scatter, and part of this scatter is in testing methods, etc. Presumably you evaluated your accuracies with the manual system and the installation system?

Mr White: We would hope that scatter has been reduced in fact by controlling the temperature rather better than under manual methods. There were some rather ancient machines in use at that time. The temperature controllers needed replacement and this was part of the history of putting in this system. I agree that you would hope to reduce scatter because you are controlling your temperature, temperature being one of the main causes of scatter in this case. We have in fact held the cost of testing down over this period. We are reducing it now, in spite of inflationary costs, by the installation of the computer.

The Chairman: Would those remarks apply equally to the tensile testing operation?

Mr White: With regard to what we have achieved on the tensile side, it is difficult to see whether we have achieved anything in the scatter of results. We probably have, because the computer is more consistent in its estimations of proof stresses and tensile strengths than operators are. Certainly the time taken to do a complete tensile test (the bulk of the time is taken in doing the graphing and calculations) has been reduced considerably by the use of a computer.

A Delegate: People have been saying that the longer term tests have been bringing the factors down. When all is said and done, we do know we could do creep tests by vibration, by heating the specimens, and by use of induction. This curve introduces vibration. We do know that many plants have been working at high temperatures where the creep figures suggest that they should not work so long without failure. A further point is that a lot of heat treatment is used to improve the creep resistance on a short-term test, but on a long-term test, surely much of the heat treatment is lost and it should really be tested in an annealed condition?

Dr Barford: We are always pleased when plant does not blow up; and, in fact, our operational experience with high-temperature plant is very good in this respect in that I do not think we have had a 'simple' creep failure in a thick section component. I use the word 'simple' to exclude the effects of welding or other onerous fabrication procedures.

Dr Mitchell: There is no point in giving it a heat treatment which will force it to break in the first ten hours.

Dr J. Congleton (University of Newcastle upon Tyne): There is an interesting point that has not been raised.

If you are concerned with scatter in creep data and accept temperature control as very important in this, you can get round the temperature control problem by using the computer to stabilize this. If you are saying that your creep properties are very much controlled by variations in temperature, you have the immediate facility for a whole range of temperature cycling which might give you more indication of why you are getting scatter rather than just trying to stabilize it. One would think that there is scope here for the mathematician to devise the right kind of cycling to give some information about the variability of data. Have there been any moves towards this?

The Chairman: There has certainly been work on thermal cycling of that type, but to my knowledge with the specific idea of determining the mechanism of creep behaviour.

Mr Johnson: The effect of temperature on scatter must not be over-emphasized and it would be quite wrong to think that scatter could be virtually eliminated if the temperature was computer-controlled. I agree entirely that it is a factor, but equally there are many other factors such as testing technique, axiality of loading, and perhaps most important, the variability of the material itself. I would be interested to know by how much the scatter has been reduced in Mr White's laboratory by the supercontrol of temperature.

Dr T. Gladman (BSC, Swinden Laboratories): On this point, I think that in much of the information that went into the parametric techniques, you will be able to discern quite clearly the difference between one cast and another cast made to the same nominal composition. If, in fact, the temperature variation caused a large proportion of the scatter, then you would not see compositional effects very clearly. I would support Mr Johnson, who indicated that the compositional variations within specification were one of the largest single factors that contribute to the scatter band. I think Mr Taylor may have seen evidence of this in his results.

Mr Johnson: I agree with you, but it is not only on the sort of macrocomposition, but also on the microcomposition, the variation and segregation and differences in the structure of the material. I think you could take two samples of identical composition-nominal composition, and still find very considerable differences. I think we might have to consider even the grain size.

Dr Barford: This has been demonstrated on a number of materials. Given reasonable temperature control during test, the cast-to-cast variations and the heat-treatment variations are clearly discernable.

ANALYSIS OF THE RESOLVED SHEAR STRESS DISTRIBUTION BENEATH A KNOOP INDENTER

C. A. Brookes, E. L. Morris, and B. Moxley

A model has recently been developed to explain the observed anisotropy in Knoop hardness measurements for a wide range of metallic and non-metallic crystals. This has shown that the nature of anisotropy is essentially determined by the primary slip systems which accommodate dislocation movement during the indentation process. Consequently, materials having the same crystal structure and slip systems should possess similar anisotropic properties. An effective resolved shear stress (τ_e') equation has been developed. The relative magnitude of the mean value of τ_e' for a Knoop indenter with its long diagonal oriented in specific crystallographic directions on a particular plane may be calculated. Thus it has been confirmed that directions which correspond to the minimum mean values of τ_e' are those of maximum hardness, and conversely so. Values of τ_e' have to be determined for each of the four indenter facets, on every possible slip system in the crystal structure, as the indenter is rotated on a given surface. Subsequently the distribution of the resolved shear stresses is established. The arithmetic mean of the maximum resolved shear stress for each facet is derived from these results and this reflects the nature of anisotropy. A computer program has now been written to calculate values of τ_e' for indentations on planes in cubic crystals and this is described.

The authors are with the Department of Engineering Science, University of Exeter
620.176

List of symbol used in the text

F axis	axis of a cylindrical element of the crystal parallel to the steepest slope of an adjacent indenter facet
H axis	axis lying in the plane of an indenter facet and parallel to the surface of indentation
SN axis	axis in the direction of the normal to the slip plane
SD axis	axis in direction of the slip direction in a slip plane
AR axis	axis of rotation in a slip plane and normal to the slip direction
α	angle of clockwise rotation of Knoop indenter

	from an initial specific crystallographic direction
τ_e'	effective resolved shear stress
P	applied force along 'F' axis
A	cross-sectional area of cylindrical element
φ	angle between 'F' axis and the normal to the slip plane
λ	angle between 'F' axis and the slip direction
ψ	angle between the axis of rotation for a given slip system and the 'H' axis
γ	angle between the 'H' axis and the slip direction in a given slip plane

Anisotropy in the hardness of crystalline solids has been firmly established.[1] This characteristic is best investigated using the Knoop indenter which consists of a diamond pyramid that produces rhombohedron-shaped indentations where the long diagonal is generally about seven times the length of the short diagonal. Hardness measurements are based only on the length of the long diagonal of the indentation when the indenter is oriented in a specific direction on a given crystallographic plane. The curve in Fig. 1a shows results obtained for indentations made with a normal load of 300g on a freshly cleaved (001) plane of magnesium oxide. The directions of maximum and minimum hardness are found when the long diagonal of the Knoop indenter is aligned in $\langle 1\bar{1}0 \rangle$ and $\langle 100 \rangle$ directions respectively. The curve in Fig. 1b is for indentations on a (110) plane. It is

generally found that the Knoop hardness of a given crystal has about the same value for a particular direction and is not significantly influenced by the plane of indentation. This can be seen by comparing the results for $\langle 100 \rangle$ and $\langle 1\bar{1}0 \rangle$ directions, on the (001) and (110) planes of magnesium oxide, shown in Fig. 1. Furthermore, materials possessing the same crystal structure and having the same slip systems show similar anisotropic properties. Thus NaCl, LiF, and MgO all have the rocksalt structure and slip on $\{110\} \langle 1\bar{1}0 \rangle$, and consequently each exhibits the same type of anisotropy.

An analysis of the indentation process based on the 'effective resolved shear stress, (τ_e')', in the bulk of the crystal beneath the indenter, has been formulated. This assumes that the indentation process, using a

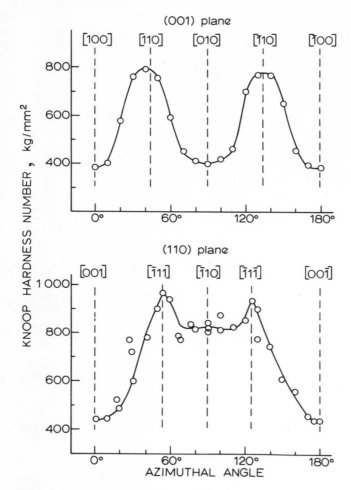

1 **Anisotropy in the hardness of magnesium oxide on the (001) and (110) planes**

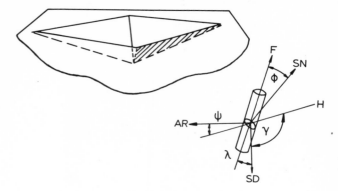

2 **Schematic illustration to show the angles $\varphi, \lambda, \psi, \gamma$ used in equation (1). The angles F : H, SD : SN, and SD : AR are all 90°**

the angle between the F axis and the normal to the slip plane, and λ is the angle between the F axis and the slip direction. The modifying function $\frac{1}{2}(\cos \psi + \sin \gamma)$, which reflects the degree of constraint imposed by the indenter facet on the rotation of the slip plane during indentation, can be determined when the axes AR and H are specified. AR is the axis, normal to SD and SN, about which the slip plane rotates during indentation. H is an axis parallel to the indenter facet and to the surface of indentation, so that ψ can now be defined as the angle between the axes AR and H. Finally, γ is the angle between the axes SD and H. The constraint term can best be demonstrated by considering the conditions where this term is either zero or unity—representing conditions of maximum and minimum restriction respectively.

Rotation of a slip plane can only take place when it is subjected to an effective resolved shear stress. Therefore the maximum constraint will be obtained when τ'_e is zero—i.e. when the whole constraint term is zero because γ is 0° and ψ is 90°. This is the situation when the axis of rotation is normal to the H axis and when SD is coincident with H. At the other extreme, the minimum constraint will be found when the axis of rotation and the H are coincident. Then the constraint term is unity because ψ is 0° and γ is 90°.

It has been suggested that the indentation process is governed by compression forces normal to the indenter facets rather than tensile forces parallel to these facets. In this case the constraint term would not change, but the Schmid Boas factor varies in certain orientations. This possibility is considered in greater detail later.

A direct correlation between the absolute hardness value and equation (1) is not possible at the present time because a representative value for A cannot be defined and the effect of work hardening on the hardness has not yet been established. Nevertheless the product of the geometric terms in the equation indicates the relative magnitude of the effective resolved shear stress for a Knoop indenter oriented in specific crystallographic directions on a particular plane.

Knoop indenter, is controlled primarily by plastic deformation, and that the value of τ'_e for a given slip system may be evaluated by the following equation:-

$$\tau'_e = \frac{P}{A} \cos \varphi \cos \lambda \ \frac{\cos \psi + \sin \gamma}{2} \ \cdots \cdots \cdots (1)$$

A vast amount of numerical data is involved in the calculation of all the values of τ'_e, for indentations on a specific plane of a particular crystal, and this is clearly best derived and handled by computer techniques.

THE EFFECTIVE RESOLVED SHEAR STRESS EQUATION

A representation of a Knoop indentation is shown in Fig. 2. For one facet, a tensile force acting along an axis parallel to the line of steepest slope of that facet (the 'F' axis) is considered. For a given slip plane in a cylindrical element of material within the bulk of the crystal, the directions of the axes SN, the normal to the slip plane, and SD, the slip direction, can be established. The first part of equation (1) can be recognized as the Schmid Boas relationship.[2] P is the applied force acting along the F axis, A is the cross-sectional area of the cylindrical element, φ is

COMPUTER APPLICATION

Stereographic projection techniques have been used to measure the angles $\varphi, \lambda, \psi, \gamma$ and thus obtain values of τ'_e, for a particular orientation of the Knoop indenter on a given plane, in the earlier work.[1] All four angles must be determined for every orientation of the indenter, for each of the four indenter facets, and all slip systems in the crystal. Thus, a considerable number of measurements and calculations are involved. These are clearly best handled by a computer, and Fortran programs have now been written for a Digital PDP8 and an ICL 4-50 computer. Only cubic crystals will be considered in this paper.

The computer calculates each value of τ'_e by first determining the unit vectors in the directions of the axes: $F, H, SD, SN,$ and AR. Cosines of the required angles are then obtained by taking scalar products of the relevant vectors. The flow chart representing the program is shown in Fig. 3. Initially, the Miller indices representing the operative slip systems, the plane of indentation, and an initial orientation of the indenter on that plane are read into the computer. All the possible slip planes in the relevant slip systems must be identified. This is done by the generation of all possible unique permutations of the three indices, while allowing for each individual index to be negated but excluding sets which are the negative of one another. The flow chart for this part of the program is shown in Fig. 4. A similar process generates all the possible directions of slip in these planes. The unit vectors in the directions of the slip plane normals and the slip directions are obtained by normalizing the Miller indices generated. To find which of the slip directions lie in particular slip planes, the scalar product of each of the slip plane normal vectors with each of the slip direction

vectors in turn is calculated. Thus are stored all the plane-direction pairs, i.e. each slip system, for which this product is zero. For each plane-direction pair, the unit vector in the direction of the axis of rotation is calculated from the vector product of the slip plane normal vector with the slip direction vector. Consequently unit vectors in the direction of $SD, SN,$ and AR for each slip system are found with respect to the cubic axes of the crystal.

We may consider that the Knoop indenter is rotating in its own set of orthogonal axes about a z axis with its long and short diagonals initially in x and y directions respectively. Then the unit vectors in the F and H directions, with respect to these axes, can be written for each of the four indenter facets for the situation in which the long diagonal is at an angle α measured clockwise from the x axis. These vectors (calculated in terms of the indenter geometry and α only) have previously been identified and are written into the program. Given the Miller indices for the plane of indentation and for an initial orientation of the long diagonal on that plane, then the axes in which the indenter is rotating are defined with respect to the axes of the crystal. A rotation matrix is therefore calculated which converts the H and F vectors, for a given value of α in the indenter axes, to vectors with respect to the crystal axes. Unit vectors in the directions of all the axes have now been determined in terms of one set of orthogonal axes. Thus values of τ'_e can be calculated and printed out for all specified values of α, all four facets, and all the slip systems generated.

RESULTS

The complete set of τ'_e curves for indentations on the (001) plane of a crystal having $\{110\}\ \langle 1\bar{1}0 \rangle$ slip systems

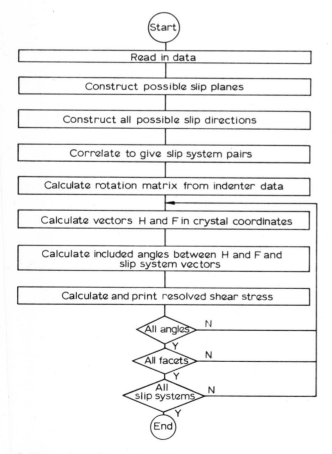

3 Flow chart for computer program

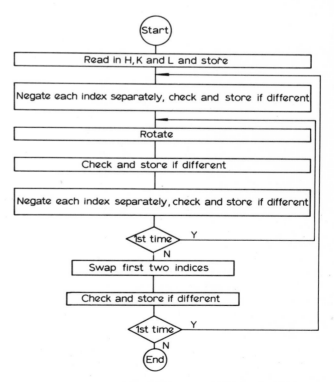

4 Flow chart for generating slip planes or slip directions

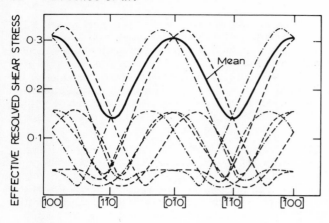

5 Complete family of τ'_e curves for (001) magnesium oxide

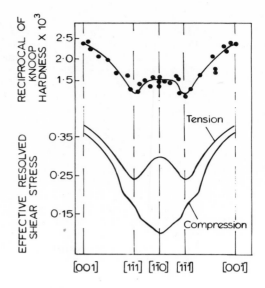

6 Comparison of analyses based on compression and tension for (110) plane of magnesium oxide

is shown in Fig. 5. The curves represent the results for each facet, each slip system and α varying from 0°–180° from an initial [100] direction. In this particular case, τ'_e for opposite facets has the same value but this is not always observed. The mean curve is obtained by taking the arithmetic mean of the maximum values of τ'_e for each of the four facets at each value of α. Comparing this mean curve with the hardness curve for (001) magnesium oxide (Fig. 1*a*) it is seen that directions corresponding to minimum values of τ'_e are the directions of maximum hardness, and conversely so. If necessary the computer program may readily be modified so that the mean values of τ'_e are calculated and printed out.

Figure 6 shows the results for the (110) plane of a crystal having $\{110\}\langle 1\bar{1}0\rangle$ slip systems. The top

curve is the reciprocal of the hardness for a (110) plane of magnesium oxide earlier shown in Fig. 1b. The curve marked 'tension' is the mean τ'_e curve obtained in the normal way using the analysis outlined earlier whilst the curve marked 'compression' is based on a similar analysis but using the concept of a compression force normal to each of the indenter facets. In most cases the resolved shear stress curves predict the same type of anisotropy whether based on tension or compression. However, it is clear that the analysis based on tension is more consistent with the measured hardness for the (110) plane of rocksalt-type crystals.

Table 1 Knoop hardness numbers for cubic crystals

Crystal	Structure	Knoop hardness number							Slip[†] system	Reference
		(001)		(110)			(111)			
		$\langle 100\rangle$	$\langle 110\rangle$	$\langle 001\rangle$	$\langle 1\bar{1}1\rangle$	$\langle 1\bar{1}0\rangle$	$\langle 1\bar{1}0\rangle$	$\langle 11\bar{2}\rangle$		
MnS	Rocksalt	122	142	119	142	142	140	140	$\{110\}\langle 1\bar{1}0\rangle$	3
MnS	"	164	183	162	194	196	162	160	"	4
LiF	"	87	93	87	97	93	93	*	"	5
LiF	"	96	103	98	120	116	*	*	"	1
NaCl	"	18	20	*	*	*	*	*	"	6
MgO	"	400	800	*	*	*	*	*	"	7
MgO	"	400	780	420	930	810	*	*	"	1
MnO	"	252	285	252	287	287	*	*	"	4
CaF$_2$	Fluorspar	178	157	*	*	*	164	158	$\{100\}\langle 011\rangle$	1
Al	fcc	23	18	23	17	18	18	*	$\{111\}\langle 1\bar{1}0\rangle$	5
Al	"	18	14	22	15	16	17	18	"	8
Ni	"	105	72	115	84	84	93	*	"	9
MnSe	"	65	43	73	54	48	57	57	"	3
Cu	"	47	34	*	*	*	*	*	"	6
C	diamond cubic	9600	6900	*	*	*	*	*	"	10
W	bcc	445	375	440	360	360	390	408	$\{1\bar{1}0\}, \{11\bar{2}\}$ $\{12\bar{3}\}, \langle 111\rangle$	11
W	"	409	337	409	343	337	337	*	"	5
Nb	"	81	59	81	63	59	59	*	"	5
Fe(Si)	"	229	183	240	196	203	198	214	"	12
V	"	103	79	97	*	78	79	92	"	13
Cr	"	139	108	159	*	115	108	123	"	13

[†] Slip systems thought to control indentation process

Table 1 gives the Knoop hardness data for a wide range of cubic crystals, both metallic and nonmetallic. It can be seen that the nature of anisotropy is governed by the crystal structure and the operative slip systems. Mean τ'_e curves have been obtained for indentations in the planes of the various cubic crystals listed in Table 1. In all cases the type of anisotropy predicted by these curves is in agreement with the hardness values listed. The computer program is currently being modified to cover other crystal structures.

ACKNOWLEDGMENTS

The authors thank Professor H. G. Edmunds for his support; the SRC and the UKAEA for grants to the Laboratory.

REFERENCES

1   C. A. Brookes *et al.*: *Proc. Roy. Soc.*, 1971, **A322**, 73

2   E. Schmid and W. Boas: 'Plasticity of crystals', 105; 1950, London, F. A. Hughes and Co.

3   E. Riewald and L. H. Van Vlack: *J. A. Ceram. Soc.* **52**, 370

4   J. W. Moore and L. H. Van Vlack: Proc. Symp. 'Anisotropy in refractory compound single crystals', 1 (eds. F. W. Vahldiek and S. A. Mersol) 220; 1968, New York, Plenum Press

5   M. Garfinkle and R. G. Garlick: *Trans. TMS-AIME*, **242**, 809

6   J. B. O'Neill: PhD Dissertation 1970, University of Bradford

7   F. P. Bowden and C. A. Brookes: 1966, *Proc. Roy. Soc.*, **A295**, 244

8   E. R. Petty: *J. Inst. Met.*, **91**, 54

9   A. L. Vecchia and W. Nicodemi: *Metallurgia Italiana*, **9**, 321

10   C. A. Brookes: Nature, 228, 5272, 660

11   G. O. Reick *et al.*: *Trans. TMS-AIME*, **242**, 575

12   F. W. Daniels and C. G. Dunn: *Trans. ASM*, **41**, 419

13   D. G. Alexander and O. N. Carlson: *Trans. Met. Soc. AIME*, **245**, 2592

DISCUSSION OF THE SECOND SESSION (part two)

In the chair: Mr J. Hewitt

Dr Mitchell: With reference to your paper, Mr Moxley, there is obviously a specific problem here as regards the program. What computer, and what language, was used?

Mr Moxley: The program used is Fortran, and it has been programmed on two computers, the PDPF digital and the ICL450. The digital takes about two hours to receive and print out the results and the ICL about three minutes, although the calculation time is the same for both.

The Chairman: The price of the computers is in about the same ratio as the time they take.

Mr W. v. d. Walls (Hoogovens): What was the time taken to develop the software for your program?

Mr Moxley: I am not too sure about the actual details of the program: that is not really the part with which I am concerned. The programs take a fortnight to run.

Dr T. B. Vaughan (IMI): I have two comments related to the earlier papers. The first is that it strikes me that Wiggins are using a computer to control a number of activities in a testing laboratory in a very imaginative way. This is the way that one would like to see computers used. Other papers so far have brought in a computer merely to do calculations, albeit rather sophisticated ones, and this is a rather obvious, application of a computer; although I would not underrate the tremendous saving in time, and even perhaps the generation of further ideas inherent in the use of a computer in a calculating mode. The second point is that from the first discussion in this session, it appeared that quite a number of people were agreeing that there was an area of ignorance concerning the factors affecting variability in creep tests. Surely Mr Greenfield would agree that this is just the sort of area where the industrial statistician could help. I do not think we would even have to propose a model. Provided enough data were made available, analysis of variance or some similar technique would no doubt help the investigators to pinpoint the various significant factors.

Mr Greenfield: There are a number of statistical techniques which can only be used by using a computer with so much data and so many variables that it would not be practicable to apply the analysis by hand. The industrial statistician would begin however, by recommending an experimental design. I know that this was discussed several times by a group of people. It has got some practical problems attached to it, has it not? You have not been able to follow the design, you have always had to use collected data rather than data resulting from the planned experiment.

Mr Taylor: Yes. I think the answer centres around the type of information collected by steelmakers and users over the years. This information was originally obtained from acceptance tests when designers wanted to know what values they could expect for a particular

material. Tests were put on which were allowed to run out to completion or were discontinued after a time if the question had been answered. There was not any actual planning introduced as a corporate arrangement until the British Steelmakers' Creep Committee was established. Work did then begin to centre upon convincing everyone that the information we had, which represented quite an appreciable investment in terms of time and money, was only useful as a general guide and that planned programs had to be introduced. The need for a planned testing program is now more fully appreciated.

The Chairman: I am not so sure that we were saying that there is a lot of scatter which is abnormal in creep data. I think that what we were saying earlier was that there was a lot of scatter within a given type of material, and Mr Johnson was saying that we know what some of the scatter is caused by. It is caused by variation in composition within the material specification, by variation in temperature, etc. What you were suggesting was that perhaps we could statistically sort out how much of the scatter was contributed by each variable.

Mr Greenfield: As I understand it so far, the variation in composition has been taken into account only in this analysis by editing rather than by including it in the analysis, so that you take specifications within certain ranges and do not include the compositional variables in the mathematical model. Of course we could start including the compositional variations in the mathematical model, but Mr Johnson raised the point of segregation within the actual testpieces. I do not know how we would treat this.

Mr Johnson: I would question whether or not there would be any real advantage, in terms of engineering design, if in fact it proved possible to eliminate scatter. In saying this, I am bearing in mind particularly that the simple laboratory tests carried out to determine the high-temperature properties yield results related to one particular set of test conditions. In many applications, the final components will not operate under such conditions, particularly as far as the stress pattern is concerned. Therefore, in any event, the laboratory test result can only serve as a guide to design.

Mr Greenfield: You may be right. Certainly the type of creep test you do does not in any way reflect the duty to which a material is put in actual service because you have got iso-stress, iso-thermal tests. You do not include any distribution of duty. The way the statistician would look at this is to day, 'I want to try various distributions', the three variants in the distribution being stress, temperature, and life, and then try to integrate out of that a marginal distribution for the lifetime and see how that actually compares with the distribution you get under constant stress and constant temperature. I imagine this would be very difficult but it is the sort of approach I would adopt.

Dr Barford: I am not quite sure what Mr Johnson means. If he means what I think he means, I am a bit shocked. Is he saying the designers are not too sure how to use the 100 000h rupture stress in the sense that they do not know precisely what stresses are operating in the complex geometries of, say, large castings because of the relaxation of internal stresses and fabrication stresses? If he is saying that steel-makers do not want to reduce the scatter because the designers do not know how to use it, that is a failure of the design which presumably will not be a per-manent thing. But if he is saying that steelmakers do not want to reduce the scatter because this would mean improving their practice, then indeed I am shocked. He is saying in effect that we are not going to get any better materials, we are going to stay with materials with a scatter of three orders of magnitude of time to rupture. I shall be very pleased if Mr Johnson will tell us which of these several interpre-tations he intended.

The Chairman: I will try and interpret what I think Mr Johnson meant. He was really saying that there is no point in identifying the cause of the variations in terms of individual segregation patterns in a test-piece, in quite a lot of the scatter we get, because, paraphrasing your argument, you are saying the variability, the segregation patterns, etc., are the sort of things you will get in the material anyway. In other words, if you are looking for a perfect-fit straight-line relationship, it is an academic exercise and not a practical one. Is that true?

Mr Johnson: Yes, I think so. In view of the major effects of inherent scatter in the material and other factors, it would simply be of academic interest to attempt to isolate the effects of small temperature variations. In this sense, the mammoth exercise in-volved could not be justified.

Dr Barford: In that case, I am shocked. Materials have enough properties without scatter being accepted an intrinsic material property. I accept that scatter arises, but what I do not accept is that nobody is pre-pared to look to see what gives rise to it.

Dr Vaughan: It is only a mammoth exercise to attempt to isolate the effects of small temperature variations, surely, if you literally have to start from scratch and obtain the data again. What I am suggesting is that data already available might possibly be analysed further in order to determine the effect of various factors.

Mr Johnson: This has been attempted, but without suc-cess. The reason for this is that in spite of the large quantity of data available, there are also a large num-ber of variables. It has proved impossible to isolate these on the basis of information to hand.

Dr Vaughan: They are not necessarily all significant. It is the job of the statistician to find out which ones are.

Dr Barford: I accept this, and with no disrespect, some of the earlier tests were rather sloppily done. The heat treatments were not closely specified; the conditions and actual mechanics of the test were per-haps not up to present-day techniques, so it has been impossible in general to separate the variables giving rise to these things, but let us not say that now, in 1971, and in the future, that we accept scatter as an inherent property.

Mr Johnson: I am sure that we will have to live with scatter for many years to come.

Dr Barford: May I ask Professor Nutting if he accepts scatter as an intrinsic property of steel?

Professor Nutting: Scatter is not an inherent charac-teristic of creep testing. In perfectly pure materials the creep test results should be reproducible. How-ever, with commercial materials where variations in composition can occur from cast to cast, then some scatter is inevitable.

Mr Johnson: The simplest form of steel is inherently heterogeneous and cannot, therefore, be compared with an element like copper.

Professor Nutting: I doubt if one could compare, because you have variables and therefore you are likely to have some scatter.

Dr Mitchell: No. May I make the point that the person using his superpure copper knows what its tensile strength or creep strength is and therefore adjusts his gauges on his testing apparatus so that there is no scatter. It is all a question of the testing conditions more than the properties of the materials themselves. I do not know how anybody can expect two people to get the same 0·2% proof stress when you consider how the 0·2% proof stress is calculated. I do not know any test department, without using a ruler and pencil, who could do it. You only get away from scatter when using superpure materials and very impure testing apparatus, and you simply adjust the testing apparatus to get the results that you know you should get.

Mr Greenfield: A lot of scatter is inherent in the ex-periment in fact, as with the Charpy transition curve, which people insist on drawing by hand because they have always done it. Then from the hand-drawn figure they try to pick off the point of inflexion, or some such point, on it. It is nonsense to do it by hand, and yet I see people everyday doing it. They will not change their ways.

Mr Johnson: Perhaps the notion that computers are capable of eliminating scatter represents one of their dangers.

Mr J. H. Woodhead (University of Sheffield): May I ask for clarification about what is being considered when the term 'scatter' is used? I find myself in a state of considerable confusion after listening to the discussion so far. Are we talking about the scatter of results obtained on testpieces from one cast of a steel or are we talking about the scatter of results obtained for a given nominal composition of steel? A designer is concerned with scatter in the latter sense. Both types of scatter are of interest to the metallur-gist but let us be quite clear about distinguishing between the two. I suspect that we ought to be con-centrating on the variability within types of steels because this is the factor which is of major impor-tance to the user.

The Chairman: I think really we are discussing the latter point about the inherent variability of a type of material.

Dr Barford: That is right. But it may not be as bad as the points on Mr Taylor's graph suggested because he had many casts on together. There is work that shows this quite clearly on casts of chromium-molybdenum-vanadium steel, in which there were six casts in various heat-treated conditions. There were perhaps two orders of magnitude scatter in rupture time in a reasonably long period, but when the separate batches were identified by both cast analysis and heat treatment, the scatter on any one curve from one cast with one heat treatment was very small. It is true that a designer is presented with a class of material from which he wants to make his bits and pieces, but he would be much happier if the scatter was less. This situation is not going to arise unless the manufacturer traces this back to the starting point to find out where the scatter comes from, and we are never going to get any further until we do this.

Professor Nutting: I have some sympathy with Dr Barford on this point. In steels it appears that the creep properties are very susceptible to the presence of trace elements. Thus it might be expected that under normal steelmaking procedures the creep properties would lie within a certain band, but if for some reason a trace element such as lead gets into the steels, abnormal behaviour may be found in creep tests but this would not immediately be attributed to Pb since this element is not usually determined in a cast analysis.

Mr Greenfield: Have these effects been estimated?

The Chairman: No. The point you raised really is that lead has been identified now but there is still a lot of discussion as to whether arsenic, zinc, antimony, and tin have equal effect; but on a general commercial question, I think Dr Barford wants an assurance that the steel industry is accepting some responsibility for examining the factors that do influence this variability or this scatter.

Mr Johnson: Such work is going on all the time. The British Steelmakers' Creep Committee is currently updating its data files and, once this is completed, the situation will be re-assessed. Experts will carry out a detailed examination of the data on each steel in turn and attempt to isolate the major variables. In particular, the aim will be to establish, and subsequently eliminate, factors which give rise to the lower portion of a particular scatter band.

The Chairman: Professor Nutting raised a point in which I am personally interested, and that is the question of deriving specifications and the implications behind this. You notice from the summary paper that I was going to give a discussion on this very aspect and I could not do so by changing jobs at a critical moment in time. But there are two things which intrigue me. First of all, the physical testing laboratories of Wiggins, and the like produce a phenomenal number of test results each week. My own Division is currently doing of the order of 2–3 000 tests per week and most of these results are filed away. Most of them are sent to customers with test release certificates. But if you take these data and run regression analyses on them and start asking questions about what causes variability within a specification, the results that invariably come out on steels are very good in the sense, for example, that the tensile properties are very closely predicted by composition. If you run regression analyses and produce results showing the regression equation for tensile strength: some constant plus a factor for carbon, silicon, manganese, etc., and produce similar equations for ductility, elongation or reduction of area, and impact properties, you then finish up with a set of equations from which you calculate the properties for any given composition. If you then look at a British Specifica-

tion and use the equations you have generated as a linear programming optimization model, you come to the conclusion that the specification and the properties do not match. This has been my experience on the seven or eight specifications we have now looked at.

It seems to me that a lot of the specifications put out by Standards Committees and by customers particularly simply duplicate what has always been know. When we come to write a specification for silicon content, we always say 15-35, do we not? It has always been that, so why should we change it. If you look at the effect of silicon on properties of simple or low-alloy steel, this is wrong, and results agreed that the silicon content should be higher than that normally quoted.

It is this question of using a computer to analyse the data and optimize properties and produce optimal results instead of average results which intrigues me, which is why I personally am very impressed by the Wiggins setup. But are we sure, in specifying residuals and trace elements and general composition, that we know what they are?

Dr Congleton: May I take up the point Professor Nutting made. I was surprised this morning, in Mr Taylor's talk, by the very wide span in results that you have. May it not be the case that you are actually trying to treat statistically a mass of data that you should not be treating as a uniform block in any way, and if you can get residuals coming in and giving you very low creep properties you should be dropping these out from your analysis. The scatter that you are really talking about is the scatter that is in average material and you should not be taking into your scatterband the things that are very specific. In other words, you should be analysing your data to say that if these results come out to what you would expect statistically, for a normal scatterband, there must be some good metallurgical reasons for this, this is not an average property. You should then be prompted to look and see why this is so very different. It is not just within the normal scatter you would expect statistically.

The Chairman: This is in fact what the Creep Committee do. The question of why they are way below the norm is what is involved, but I do not think you would accept for a moment that because some results are lowish you should miss them out of the calculations. You would not accept that, would you?

Dr Congleton: No.

Mr Johnson: If results are very low, an examination of the history of the cast concerned will generally provide an explanation. If the performance of a cast is not typical, then it is wrong to include it in a statistical analysis aimed at deriving typical properties. Data should always be technically appraised from this point of view and screened before it is fed to a computer. This is not to say that all low points (or indeed high points) should automatically be excluded; they should simply be questioned and then eliminated if there is sound technical justification for their omission.

Mr Greenfield: If you have a good technical reason for excluding it; but only if you do find a good technical reason.

Dr Duckworth: Do you then also exclude other results which are in the same group? If, for example, both points were excluded because they were heat treated by the same firm whose heat-treatment practice was suspect, would you be certain to remove all results of specimens heat treated by that same firm?

Mr Johnson: Generally speaking, yes. Such results should be removed, even though they happen to lie within the typical scatterband.

A Delegate: Two years ago a paper appeared in the American Welding Journal entitled 'Why certain heats of stainless steel will not weld' and this was about an automatic welding setup which had been running using argon-arc welding; certain parameters were being fed into this system, and it was found over the course of a number of years that certain components welded up excellently, but for no apparent reasons, other components would not. The systems were monitored and the conditions were maintained as reasonably identical as possible: the people responsible for running this laboratory had an analysis of the heats available, and they did a data logging operation.

This report purported to explain why certain heats were not weldable. They analysed the data, with composition and welding performance in an undefined manner which has not been explained in the report, but the upshot was that the unweldable or poorly-weldable material seemed to have, if my recollection is correct, a silicon : manganese ratio which varied from the good welding and material, and they rationalized this into some sort of theory that if the silicon : manganese ratio was in this particular value, they had a very fluid material which therefore was not so weldable.

I tell this tale because I think that with these 200 tensile results that Laboratory *A* is generating per week, for laboratory *B* there is a big opportunity for data logging and relating to the analysis of the heats, so that this sort of analysis-of-properties-versus-composition can be elicited. I think the problem of scatter is not so much that the designer is bothered with scatter as such. If he knows the scatter then he can define the extreme value for the level of reliability he wants.

Dr Congleton: I would like to take up an earlier point about putting in rogue specimens: it is that the rogue point should not be forgotten. It should not be used in the mathematical analysis, but there must be some very good reason why it exists. Science will not be advanced if it is completely ignored. The statistician should be looking at the data to highlight rogue points

and draw attention to them as well as giving a meaningful statement about the bulk of the results.

Dr Duckworth: A computer would determine as significant a correlation coefficient when the data, plotted out, actually showed two distinct population clusters, especially if these were at some distance from each other. It is very important if spatial relationships are being considered that the computer should plot the results out as well as computing their parameters. It is important, therefore, that the physical metallurgist should know sufficient about the statistician's art to be able to have some appreciation of what the statistician is doing and not just accept the results of t tests, F tests, etc.

I have, as many of you know, developed a series of simple statistical techniques which physical metallurgists can apply to their data before submitting them to full computation.* The need for these was demonstrated to me very clearly many years ago when I was asked to assist an engineer to plan an experiment on the flatness of thrust washers for engine bearings. These were stamped out of steel strip and one of the variables the engineer was interested in was the relationship of the final thrust washer flatness to its original position in the steel strip. After carrying out the experiment he found to his amazement when he analysed the data himself using some of these simple statistical techniques, that the flattest washers came from the tightly wrapped inside of the coil. If he had not seen for himself from his analysis that this was so he would have rejected this conclusion had it been produced by a computer and used it as yet another example of the absurd fallibility of statistical techniques. Because he had to accept his own findings he worried about it for several nights until he finally remembered that he had had to pre-flatten the interior of the coil before he fed it through the thrust washer machine. It is therefore important to look at the data yourself using simple techniques before passing them over to a statistician.

Dr Mitchell: Mr Greenfield's interactive programmes could really be a solution to this, where the metallurgist is able, without having to count higher than ten, to use some very sophisticated techniques, and where, without being able to say the statistician is a fool because he says the inside one is the flattest, he would have to feel it was himself who was getting the wrong answer and try to work out for himself why the answers were wrong.

*W. E. Duckworth: 'Statistical techniques in technological research'; 1968, London, Methuen

APPLICATION OF QUANTITATIVE METALLOGRAPHIC TECHNIQUES TO THE STUDY OF OSTWALD RIPENING DURING THE TEMPERING OF STEELS

E. Gilbart, T. A. Hughes, and J. Nutting

The introduction of an etch depth correction into a Scheil-type analysis in converting two-dimensional section distributions to distributions of spheres can account for large overestimates in the apparent volume fraction. In the Fe-Fe$_3$C system, realistic values of etch depth yield volume fractions of carbide which are more in accord with the carbon analysis than those reported previously. However, the appearance of negative numbers of particles in the smallest size range has not been completely eliminated, and accumulated grouping errors may well be significant in this region. The importance of grouping errors in this type of analysis is not well understood and it is hoped that a systematic approach using a computer on ideal particle distributions may yield some useful information. This programme has formed part of a much wider investigation into the kinetics of particle coarsening during the tempering of steels. The results of this work will include the etch depth correction, and will be reported elsewhere.

The authors are in the Department of Metallurgy, University of Leeds   621.785.7

Much has been written about the morphological changes occurring during the decomposition of super-saturated solid solutions, and from these studies many attempts have been made at developing quantitative relationships between morphological parameters and specific mechanical properties. Further developments of this approach are necessary, but progress is slow since difficulties are often encountered in applying quantitative methods to the determination of true microstructural parameters. The first basic problem relates to the transformation of measurements on a two-dimensional plane of section into three-dimensional spatial distributions from which the true parameters can then be calculated. These difficulties are not overcome even in thin foil transmission microscopy, since the specimen thickness is often small in relation to the features of interest, variable, and difficult to measure.

Assuming that a random section can be taken through a specimen (an assumption which is frequently not valid) the methods of analysis on the plane section are derived from the simply proved fact that

$$V_f = \frac{\Sigma v}{V} = \frac{\Sigma a}{A} = \frac{\Sigma i}{I} \quad \dots\dots\dots\dots\dots\dots (1)$$

where

V_f = volume fraction of the second phase
v = volume of each particle
V = total volume
a = area of each particle on a random plane
A = total area
i = intercept length of a random line in each particle
I = total length of line

Many subsequent arguments can be developed from this expression and these are presented in detail by Underwood.[1]

The chief quantitative problem in the metallography of two or multiphase alloys is to determine the true particle size distributions from the observed size distributions on the plane of section. In the case of spherical particles dispersed in a matrix, all planar sections will be circular, and from the circle size distribution it is possible to determine the sphere size distribution by a variety of methods.[2,3,4] When the particles are not spherical certain assumptions have to be made about their shape. If the particles are ellipsoidal and maintain a fixed-aspect ratio independent of size, then again by suitable analysis it is possible to obtain the true particle size distribution.[5]

From true particle size distributions it is possible to calculate various parameters such as interfacial area/unit volume, centre-centre spacing and surface-surface spacing, while from a detailed analysis of size distributions as a function of other metallurgical parameters such as time and temperature of aging treatments it is possible to determine growth behaviour of particular size fractions.[6]

However, in most metallographic techniques it is almost essential that some form of etching is used to reveal structural features, chiefly by differential chemical dissolution of the phases to produce a relief effect. When the optical microscope was the chief method of structure determination, the influence of etch depth upon structural parameters was never seriously considered, since the etch depths used were of the same order of magnitude as the resolving power of the microscope, and usually significant measurements were made of features many times larger than the resolution distance. With the introduction of electron microscopes, metallographic techniques were now available in which the resolution was very much smaller than the minimum etch depth that could be used, and as a consequence the analytical methods for determining true particle size distributions and volume fractions could no longer be applied without modification. The first attempt at applying corrections for depth of etching was made by Hyam and Nutting,[6] when using Formvar replicas in the electron microscope to study tempering reactions in plain carbon steels.

The difficulties of the method developed by Hyam and Nutting were associated with the tedious computation, and to simplify this certain assumptions had to be made about the constancy of the etch depth. As a consequence of this and other simplifications, the apparent volume fractions of dispersed carbides in tempered steels were considerably greater than those to be expected from the carbon analyses.

Since this earlier work there have been considerable improvements in electron microscopes, replica techniques, and in the availability of digital computers. All these have enabled more sophisticated analytical methods to be applied in determining true carbide particle size distributions in tempered low-carbon steels, and these are described below.

TEMPERING TREATMENTS AND SPECIMEN PREPARATION

In order to provide a structure which would show an idealized distribution of spherical carbide particles within ferrite grains and a minimum of intergranular precipitation, an 0·24%C high-purity iron alloy was used. After very fast quenching from 950°C, each specimen was cold-rolled 45% and then flash tempered for 5 min at 600°C. The specimens were then rolled to give a 20% reduction of area and retem-

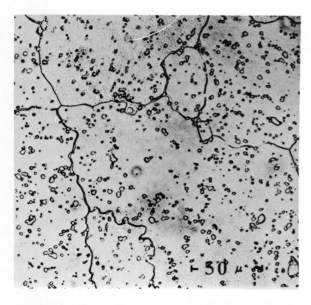

1 Optical micrograph of a 0·24%C steel, quenched and tempered for 20h at 700°C

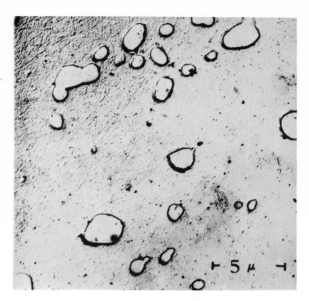

2 Platinum-shadowed carbon replica of the steel in Fig. 1

pered for differing times at 700°C so producing large ferrite grains having diameters of up to 150 μm.

The whole of the specimen-polishing procedure was designed to avoid developing relief effects, and it was found that to achieve this water-containing fluids could not be used. The initial polishing was with diamond in a mixture of paraffin and carbon tetrachloride, and the final polish was obtained with cerium oxide in ethylene glycol. A fine napped microcloth is essential for the final stage of polishing. The etchant used was 2% nital in methanol and this was applied by swabbing at −50°C. The requirements were to reveal the particles and delineate the recrystallized ferrite grain boundaries developed during the final temper. The swabbing procedure helped to remove the carbide particles loosened from the surface by dissolution of the ferrite. With commercial materials less elaborate etching procedures could be used, since when impurities are present the ferrite grain boundaries are readily delineated.

Direct carbon replicas were prepared in the usual way and these were then shadowed with Pt at a large angle to accentuate the carbide-ferrite interface. The specimens were then examined in the Philips EM300 electron microscope. On each occasion of use the magnification of the microscope was determined so that strict comparability could be maintained between the different micrographs used for analysis.

An enlargement of each photomicrograph was obtained and this print was used for particle size determination. The total magnifications on the prints ranged from 3 000–10 000 depending upon the size ranges of the structures being examined. Sizing of the particles was carried out on a Zeiss Particle Size Analyser (TGZ 3) using an arithmetic progression of size intervals. The particles were sized into 30–40 groups within the range of 1–20 mms on the prints. Subsequently these groups were combined to give between 10 and 15 groups to which the statistical analysis was subsequently applied. Typical microstructures used for analysis are shown in Figs. 1 and 2.

DETERMINATION OF TRUE SPHERE SIZE DISTRIBUTIONS FROM OBSERVED SECTIONS

The majority of analytical techniques for obtaining sphere size distributions from two-dimensional planar sections are based on the manipulation of experimentally determined distributions of circle diameters. Other methods, involving area[7] and chord length[8] distributions, have been developed but have not been applied as extensively as that based on diameters. In each case the primary object is to characterize the sphere distribution in terms of a mean diameter, a standard deviation, and also the number of spheres per unit volume. Additional parameters such as volume fraction and mean particle spacing may be simply computed from the true distribution.

Several techniques are available to convert experimental section diameters into true sphere diameters and most of these are founded on an original method due to Scheil.[2] Subsequent rationalization and correction by Schwartz,[3] and further modification by Saltykov,[4] did not improve on the accuracy of the Scheil technique but served only to simplify the mechanics of computation in an era when computers were neither speedy nor electronic. In the event, later modifications to the method remove all possibility of retaining simple, general equations, and the real advantage of the Saltykov method is lost. Thus, in the analyses to be presented later in this text the fundamental equations are those of Scheil, but minor alterations are made and inconsistencies removed without giving specific details.

Before proceeding to the extension of existing techniques for the conversion of section distributions, it is instructive to develop the basic statistical ideas in simple cases and then to extend these to more complex practical systems.

Distribution of uniform spheres

When a random section is taken through an opaque body containing a dispersion of uniform spheres, a continuous distribution of circular sections will be observed ranging in diameter from zero to that of the spheres themselves. In this case, the number of spheres per unit volume, N_V, is related to the total observed number of sections per unit area, N_A, by the simple relation:

$$N_V = \frac{N_A}{D} \dots\dots\dots\dots\dots\dots(2)$$

where D = diameter of the spheres.

This equation derives from the fact that only those spheres whose centres lie within a vertical distance $\pm D/2$ of the section plane will be intersected by it. A general feature of all the methods to be considered is that the distribution of circle diameters is not treated as a continuous function but is divided up into discrete size intervals. Thus, if only the number of sections of a particular size $N_A(i)$, characterized by a diameter between d_{i-1} and d_i is counted (i is an integer defining the particular section group) then

$$N_V = \frac{N_A(i)}{D} \cdot \frac{1}{P(i)} \dots\dots\dots\dots\dots(3)$$

where $P(i)$ = probability of the plane intersecting the spheres to produce sections in the size range d_{i-i}-d_i. This type of probability term appears consistently throughout the Scheil type analysis and may be calculated using simple geometrical considerations. In the general case, $P(i,j)$, the probability of sectioning a sphere of diameter D_j to produce sections in the range d_{i-i} to d_i is given by

$$P(i,j) = \frac{1}{D_j}\left[\sqrt{d_j^2 - d_{i-1}^2} - \sqrt{d_j^2 - d_i^2}\right]\dots\dots(4)$$

Dispersion of non-uniform spheres

A random section through an opaque body containing a distribution of spheres of different diameters will again produce a range of section diameters varying from zero to that of the maximum sphere diameter. However, in this case the observed number of sections in size ranges smaller than the maximum is derived not only from spheres of the same diameter but also from non-equatorial sectioning of all spheres from larger size groups. This may be expressed mathematically in the following form:

$$N_A(i) = \sum_{j=i}^{j=k} N_A(i,j) \dots\dots\dots\dots\dots (5)$$

where

$N_A(i)$ = observed number of sections per unit area in group i
$N_A(i,j)$ = number of sections per unit area in group i, produced by sectioning spheres in group j
k = total number of groups.

For example, if the sections are sized into four groups, then the observed number in each group, $N_A(i)$, will be made up in the following way:

(i) $N_A(1) = N_A(1,1) + N_A(1,2) + N_A(1,3) + N_A(1,4)$
(ii) $N_A(2) = N_A(2,2) + N_A(2,3) + N_A(2,4)$
(iii) $N_A(3) = N_A(3,3) + N_A(3,4)$
(iv) $N_A(4) = N_A(4,4) \dots\dots\dots\dots\dots(6)$

Since $N_A(4)$ is a measured quantity, then $N_A(4,4)$ is known and this may be converted to $N_V(4)$ using equation (3):

$$N_V(4) = \frac{N_A(4)}{D(4)} \frac{1}{P(4,4)}$$

Having calculated $N_V(4)$, it is now possible to derive $N(3,4)$, the contribution of spheres in group 4 to sections in group 3:

$$N_A(3,4) = N_V(4) \, D(4) \, P(3,4)$$

Hence the required value of $N_A(3,3)$ may be found by subtracting the calculated value of $N_A(3,4)$ from the observed value of $N_A(3)$.
Then

$$N_V(3) = \frac{N_A(3,3)}{D(3)} \frac{1}{P(3,3)}$$

and

$$N_V(3) = \frac{N_A(3) - N_V(4) \, D(4) \, P(3,4)}{D(3) \, P(3,3)}$$

This process is continued by repeatedly subtracting the contributions of larger spheres to smaller section groups down to the smallest size group. Thus

$$N_V(2) = \frac{N_A(2) - N_V(4) \, D(4) \, P(2, 4) - N_V(3) \, D(3) \, P(2, 3)}{D(2) \, P(2, 2)}$$

The main disadvantage of the method now becomes apparent, in that errors introduced into the calculation are accumulated and become more significant as the particle size decreases.

In practice, the mechanics of the calculation can be simplified by treating the terms $D(j) \, P(i, j)$ as a series of numerical coefficients. Scheil's original technique used coefficients of this form but suffered from the disadvantage that the values of the coefficients changed when the total number of groups was changed. By introducing a parameter, Δ, defined as the ratio of the maximum sphere diameter to the total number of groups, Saltykov removed this limitation and produced a table of precalculated coefficients suitable for use with any number of size intervals in an arithmetic progression.

An increasing simplification of this general approach was provided by Hyam and Nutting, who used size intervals increasing in a $\sqrt{2}$ geometrical progression. Under these conditions the number of coefficients was drastically reduced and became equal to the total number of size groups. However, this also led to very large class intervals for the larger particle sizes, and the method is only suitable in cases where the distribution is log-normal. Although this type of distribution occurs frequently in practical systems, grouping errors in the larger size classes are increased, and are accumulated in subsequent calculations. The errors become particularly apparent when determining the number of spheres in the smaller size ranges.

DETERMINATION OF TRUE SPHERE SIZE DISTRIBUTIONS FROM OBSERVED SECTIONS ON AN ETCHED SURFACE

Previous experimental evidence from the determination of particle size distribution in the Fe-Fe$_3$C system has shown that consistent discrepancies appear, which cannot be adequately accounted for.[6,9] In particular, the observed carbide volume fractions calculated from both area fractions and from corrected sphere size distributions were always much larger than those expected from the carbon content of the steel. However, the discrepancy decreased as the mean particle size increased during coarsening produced by prolonged tempering. Since the error was large even in high-purity materials, explanations cannot be based on non-stoichiometric compositions or the formation of mixed alloy carbides. Another unusual feature of the corrected distributions was that not infrequently the smallest size group apparently contained a negative number of spheres. In some cases this could be accounted for by experimental limitations in resolving sections in the smallest size group. Subsequently, it became apparent that when the etch depth on the specimen surface was of the same order as the sphere diameters the observed circle distribution would be distorted. In particular,

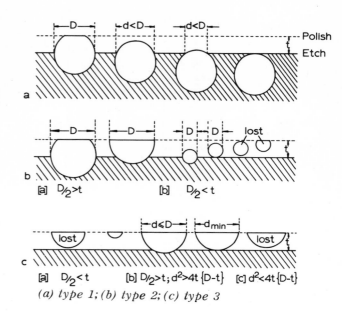

(a) type 1; (b) type 2; (c) type 3

3 Three types of sphere

this would be expected to increase the apparent volume fraction because observations are made not on a two-dimensional plane but in a three-dimensional volume in which more particles are revealed by the etching process. In the only other attempt to account for the effect of etch depth, Hyam and Nutting pointed out that in the normal case where particles are unattacked and left proud of the surround matrix, the probability of observing small sections would be decreased and that of observing large sections increased. The ensuing theory did not however consider all the possibilities of particle size and position which would be affected by etching.

The development outlined below is a more systematic attempt to account quantitatively for etching effects in the normal case where only the matrix is attacked. In order to predict this effect, it is necessary to classify the spheres according to the positions of their centres relative to the plane of polish and the new 'etch' plane.

Type 1 spheres—centres below the new etch plane

Particles of this type will produce the expected distribution of section sizes ranging from zero to those of the maximum sphere size, provided that their centres lie within a distance $D/2$ from the etch plane. Fig. 3a.

Type 2 spheres—centres between the plane of polish and the new etch plane

Spheres of this type must be further subdivided according to whether they have radii larger or smaller than the etch depth.

(i) $D/2 > t$ (t = etch depth). All spheres of this type will be retained and will appear as sections with the true sphere diameter

(ii) $D/2 < t$. The only spheres of this type which will be retained are those whose centres lie within a distance ($D/2$) above the etch plane. These will again appear as sections with the true sphere diameter. Fig. 3b.

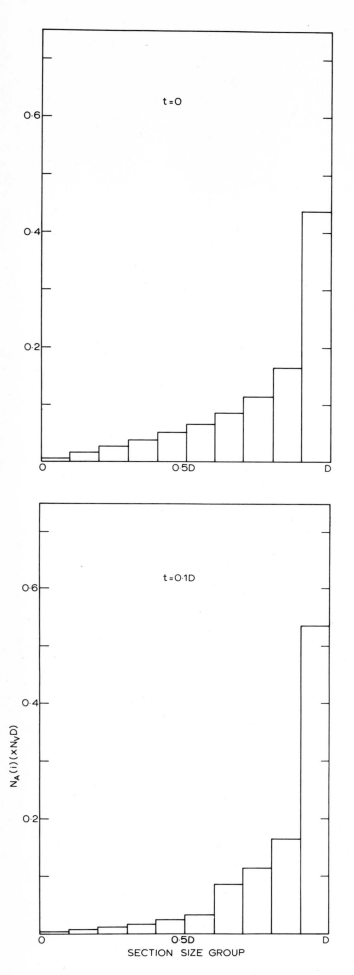

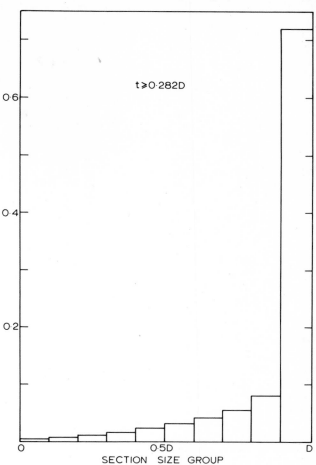

4 Effect of etch depth on the distribution of sections in ten equi-sized groups from a dispersion of uniform spheres

Type 3 spheres—centres above the etch plane

The first restriction on this type of sphere is that only those with $D/2 > t$ can intersect the new etch plane. Furthermore, the observed section size is that on the original plane of polish and this has a minimum value depending on the sphere diameter, etch depth, and the height of the centre above the etch plane. It may be shown that the minimum section size contributed by spheres in this group is given by $d^2 = 4t(D-t)$, Fig. 3c.

In order to demonstrate the method of estimating the number of spheres per unit volume, it is again useful to consider the simple case of a distribution of uniform spheres in a solid body.

The total number of sections per unit area $N_A(i)$ in a given size range d_{i-1} to d_i on an etched plane may be obtained by summing the individual contributions of each type of sphere in turn. If the diameter of the spheres is D and the etch depth t, then the contributions from each type are:

Type 1 $N_V \dfrac{D}{2} P(i)$

Type 2 $N_V t$ if $D/2 \geqslant t$
and $D = d_i$

$N_V \dfrac{D}{2}$ if $D/2 < t$
and $D = d_i$

0 if $D \neq d_i$

Type 3 $N_V \dfrac{D}{2} P(i)$ if $D/2 > t$
and $d_i^2 > 4t\,(D-t)$

0 if $D/2 \leqslant t$
or $d_i^2 \leqslant 4t\,(D-t)$

Thus the total value of $N_A(i)$ subject to the conditions imposed on each sphere type is

$$N_A(i) = N_V \frac{D}{2} P(i) + [N_V t] \text{ or } \left[N_V \frac{D}{2}\right] \text{ or } [0]$$

$$+ \left[N_V \frac{D}{2} P(i)\right] \text{ or } [0] \quad \ldots \ldots \ldots \ldots \ldots (7)$$

Then

$$N_V = \frac{N_A(i)}{\dfrac{D}{2} P(i) + [t] \text{ or } [D/2] \text{ or } [0] + \left[\dfrac{D}{2} P(i)\right] \text{ or } [0]} \quad (8)$$

When $t = 0$, $N_V = \dfrac{N_A(i)}{D\,P(i)}$, which is equation (2)

As would be expected from a purely qualitative assessment of the effect of etch depth, the greatest change in the observed distribution of sections occurs in the largest size group. Figure 4 shows how the values of $N_A(i)$, expressed in terms of N_V and D, change for each of ten equal size groups as the etch depth is increased. The values of $N_A(i)$ remain constant or decrease with increasing etch depth in all but the largest group where it increases to a maximum at $t \geqslant 0.282\,D$. Thus increasing the etch depth beyond about one-third of the sphere diameter does not produce any further change in the observed distribution of sections. This effect is also apparent in the variation of observed area fraction with etch depth, as shown in Fig. 5. In this calculation a discontinuous distribution of section sizes is assumed, and grouping errors lead to a discrepancy in the calculated area fraction. The area contributed by each size group is based on the area of a similar number of sections with a diameter equal to the mean group diameter. This leads to a calculated area fraction

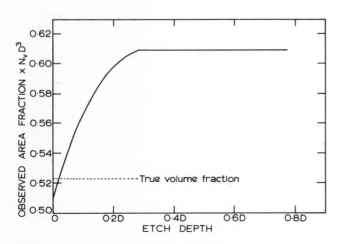

5 Effect of etch depth on the observed area fraction of a dispersion of uniform spheres

at zero etch depth of $0.510 N_V D^3$, whereas the true volume fraction is $\dfrac{4\pi}{3}\,\dfrac{N_V D^3}{8} = 0.523 N_V D^3$, that is to say an error of $\sim 2\tfrac{1}{2}\%$. Extension of the preceding theory to account for etch depth in systems containing a dispersion of non-uniform spheres is arithmetically complex. However, it is identical in essence to the repeated subtraction method of Scheil, except that a completely new array of numerical coefficients, derived from the denominator of equation (8), must be evaluated whenever the number of groups or the etch depth is changed. Since it is unlikely that both of these will remain unaltered, it is almost essential to make use of a high-speed digital computer for the analysis. In the present work an English Electric KDF9 computer was used.

THE COMPUTER PROGRAM

The basic computer program is given in the block diagram, Fig. 6. The program was written in Algol, and access to the KDF-9 was via a multiple remote job entry system developed by Leeds University Computing Laboratory. This uses on-line teletype terminals operating through a subsidiary PDP8 computer which provides facilities for the establishment, amendment, and running of programs stored as 'files' on the KDF-9 disk store. The input data are the measured circle size distributions, and from these are obtained the values of $N_A(i)$. These are then summed to give the various areal parameters. The values of the correction coefficients for the different size groups are then calculated for a given

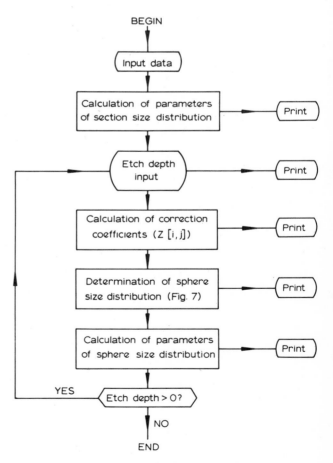

6 Block diagram of the computer program

value of the etch depth and these are stored for subsequent use. The presence of conditional branching statements introduced by the etch depth correction is dealt with by the Algol 'if clause'. The true sphere size distribution is then determined and the details of this part of the program are given in the flow chart, Fig. 7. The major part of the program may be repeated for other etch depths over a range of physically meaningful values. The whole calculation takes about twelve seconds.

RESULTS AND DISCUSSION

Initial experiments have been carried out on a sample of the steel tempered for 20h at 700°C. The observed area distribution and corrected sphere size distributions at three etch depths are shown in Fig. 8. Arithmetic mean diameters and harmonic mean diameters of the carbide particles as a function of etch depth are given in Fig. 9. (The harmonic mean diameter is included here because it is related to surface area and the driving force for particle coarsening stems from a reduction in interfacial free energy and thus surface area). Further analysis of the corrected distributions indicates that they conform approximately to a log normal distribution law.

It is apparent from these results that the general effect of increasing the etch depth is to decrease the

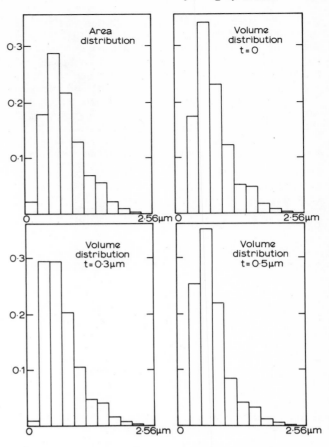

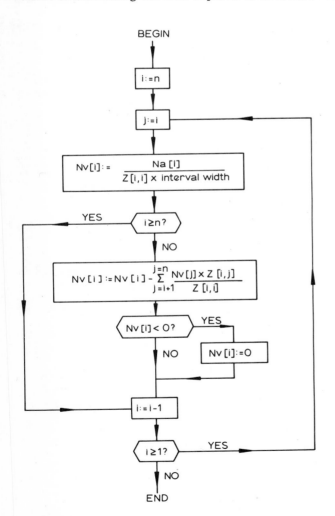

7 Flow chart of the conversion from section to sphere size distribution

8 Observed area distribution and corrected sphere size distributions at three values of etch depth (all distributions are normalized)

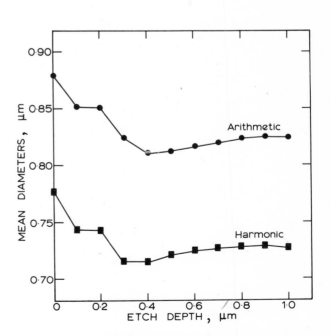

9 Effect of etch depth on the calculated mean diameters of carbide in the tempered steel

calculated mean diameter, although there is a minimum in each of the curves of Fig. 9 at etch depths of about 0·4 μm. A similar trend is apparent in both the standard deviations and the total number of particles per unit volume. However, the changes in all these parameters are generally small and are not usually greater than about 10%. A further feature is that negative numbers of spheres are predicted in the smallest size group at etch depths of zero and 0·5 μm with a positive number at 0·3 μm. This discrepancy is probably due to grouping errors.

From the corrected particle size distributions it is possible to calculate the volume fractions of the carbide, and the variation of this with the etch depth is shown in Fig. 10. Assuming that 0·02%C is retained in saturated solid solution at 700°C, then the expected volume fraction of stoichiometric Fe_3C in a 0·24% C steel would be 3·55%. As can be seen, this volume fraction may be obtained from the observed size distribution at an etch depth of about 0·25-0·30 μm, and measurements on the replicas indicate that the experimental etch depths were about this value. An important feature of Fig. 10 is that changes in volume fraction produced by realistic changes in etch depth are much larger (~35%) than the corresponding variations in mean diameter. Thus it is very possible that some of the discrepancies reported by Hyam and Nutting could arise from their incorrect assessment of the effects of etch depth.

Other possible sources of error in this type of analysis are numerous, but the significance of many may be reduced to an unimportant level by careful control of the experimental techniques. Failure to resolve very small sections will not radically affect either the mean sphere diameter or volume fraction provided that the peak in the distribution curve does not lie in or near the smallest group. It may, however, produce negative numbers of spheres in the first group. A further source of error in detecting small particles may originate from the etching process, whereby loosely bonded spheres are removed by preferential attack at the particle-matrix interface. However, there was no experimental evidence for the loss of larger particles by this mechanism.

Orientation dependent etching rates in the matrix, which would result in etch depth variations from grain to grain, were not detected on plastic replicas as sudden changes in thickness, and the initial assumption of constant etch depth appears to be correct.

Assumptions as to the true spherical nature of the carbides appear to be justified by the shapes of the sections observed. However, a complicating factor was provided by closely spaced particles which coalesced, and these were sized as separate entities with the diameter of an equivalent circular section. There was no experimental evidence for the presence of polyhedral particles although replica techniques on deeply etched surfaces may not be sufficiently sensitive to detect them. Further transmission electron microscopy may well be necessary to ascertain the exact shapes of the carbides.

Probably the most important source of error in the analysis is due to the assumption of discontinuous distributions of particle sizes in a finite number of groups. Some authors[2,10] have shown that volume parameters such as V_f and N_V increase as the number of groups is increased. This observations was borne out in the present investigation, and indicates that comparison of experimental data should be based on distributions containing a similar number of groups. Thus as the mean particle size increases the sizing range of the groups should be increased.

ACKNOWLEDGMENT

One of us (EG) is grateful to the Scientific Research Council for financial support during the course of the investigation.

REFERENCES

1   E. E. Underwood: 'Quantitative Stereology'; 1970, Addison-Wesley
2   E. Scheil: *Metallkunde*, 1935, **27**, 199
3   H. A. Schwartz: *Metals and Alloys*, 1934, **5**, 139
4   S. A. Saltykov: *Stereometric Metallography*, 2nd Edition 1958, Moscow: Metallurgizdat
5   R. T. DeHoff and F. N. Rhines: *Trans. AIME*, 1961, **221**, 975
6   E. D. Hyam and J. Nutting: *JISI*, 1956, **184**, 148
7   W. A. Johnson: *Metal Progress*, 1946, **49**, 87
8   J. W. Cahn and R. L. Fullman: *Trans AIME*, 1956, **206**, 610
9   O. Bannyh *et al.*: *Jernkont. Ann.*, 1962, **146**, 714
10   H. B. Aaron *et al.*: Paper 16, Proc. First Int Cong. for Stereology, Vienna, 1963

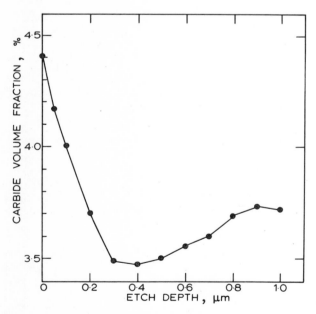

10 Effect of etch depth on the calculated volume fraction

QUANTITATIVE INFORMATION FROM ELECTRON MICROSCOPE IMAGES OF CRYSTAL DEFECTS

M. J. Whelan

The theory of diffraction contrast of transmission electron microscope images of perfect crystals and crystals containing defects is reviewed. Examples are given of the application of the theory to the computation of images of stacking faults and dislocations, and to problems of interpretation of the contrast in terms of defect models. The method of computer-simulation of images is also reviewed. Finally, mention is made of new high-resolution electron microscope techniques of use in studying defect geometry.

The author is in the Department of Metallurgy, University of Oxford

620. 187. 22:548. 73:681. 177. 8

Nowadays digital computers are used extensively in the development of techniques of interpretation of electron microscope images of crystalline materials containing lattice defects such as dislocations, stacking faults, precipitates, point defect clusters, and so on. The fundamental problem is concerned with obtaining information from the contrast effects visible on the images. For example we may want to know the Burgers vector of a dislocation, or the nature (vacancy or interstitial?) of point defect clusters produced say by radiation damage, or perhaps the shear vector of a stacking fault. That such information can be obtained by careful experiment and interpretation testifies to the power of the transmission electron microscope method as a tool in the study of defects.

First, we must have a theory of contrast effects, and the salient points of the theory are briefly outlined. Second, we have to solve the equations of the theory for particular defect strain fields. The solution of the equations by analytical methods is usually intractable, and hence resort must be made to solution by numerical integration, and for this the digital computer is indispensable. Finally, the digital computer is also used in other ways, such as in the pictorial display of the calculated image, or in otherwise intractable calculations of the geometry of crystal defects from elasticity theory, from which other useful information such as fault energies may be derived.

THEORETICAL ASPECTS OF THE PROBLEM

The situation obtained in the examination of thin crystals (foils) in the transmission electron microscope is illustrated in Figs. 1 and 2. The incident electron wave is diffracted by the crystal structure, so that a transmitted and (in general) several diffracted beams emerge from the bottom of the foil. The image is usually formed by selecting only one of these beams with the objective lens aperture, giving rise to bright-field and dark-field images as illustrated. Often only one strong diffracted beam is excited and this is referred to as the two-beam situation. By tilting the crystal away from the Bragg reflecting position or by using a very thin foil, the diffracted beam can be made very weak, so that the emerging transmitted beam amplitude ϕ_0 differs only slightly from unity (the incident amplitude). This is known as a kinematical condition. The imaging lenses of the electron microscope need not concern us in detail. We note only that it is possible to energize the lenses to form both images of the specimen and diffraction patterns from small selected areas. The image can therefore be correlated with the crystallography as derived from the diffraction pattern. In particular the method is used to determine which Bragg reflections are excited in a particular region.

The geometry of Bragg reflection is shown in Fig. 2 by means of the Ewald sphere construction. The reciprocal lattice vector of the Bragg reflection is denoted by g, and the deviation from the exact Bragg condition by s. Experimentally s may be varied by tilting the specimen or the incident beam. The electron wavelength λ at 100 keV is about $\frac{1}{25}$ Å. In the kinematical theory, the atoms are at lattice positions r_n and the scattered amplitude is proportional to

$$\sum_n \exp(2\pi i(g + s)r_n) = \sum_n \exp(2\pi i s z_n) \ldots \ldots (1)$$

where the coordinate z_n is normal to the foil surface. This sum can be approximated by an integral and the scattered amplitude is proportional to

$$\frac{\sin \pi t s}{\pi s} \ldots \ldots \ldots \ldots \ldots \ldots \ldots \ldots \ldots \ldots \ldots \ldots (2)$$

where t is the foil thickness (Fig. 1). The intensity distribution given by equation (2) is in the form of a spike through g perpendicular to the foil surface. The intensity along the spike is as shown in Fig. 3a.

The width of the spike in Fig. 3a depends on t, and hence as t varies (e.g. as in a wedge crystal), the intensity of the image will oscillate as shown in Fig. 4a provided s is not zero. The periodicity of the oscillation in depth is s^{-1}. Near the Bragg position the kinematical theory cannot be used because the diffracted wave becomes strong. The situation is then as illustrated schematically in Figs. 3b and 4b. The depth periodicity (Fig. 4b) then becomes a maximum, known as the extinction distance ξ_g. ξ_g is usually in the range 200 to 600 Å for low order reflections in metals. Fringes characteristic of thickness variation or bending of the foil are well known (Fig. 5a, b); similar fringes occur at inclined planar defects such as stacking faults (Fig. 9).

Near a crystal defect an atom is displaced by R_n from its perfect crystal position r_n. Equation (1) is then replaced by

$$\sum_n \exp(2\pi i(g + s) \cdot (r_n + R_n))$$
$$\simeq \sum_n \exp(2\pi i s z_n) \exp(2\pi i g.R_n) \cdots\cdots\cdots (3)$$

The quantity $2\pi g.R_n$ is usually referred to as the phase angle α. Equation (3) shows that the scattering near a defect is different from that in the perfect crystal because of the additional phase term α due

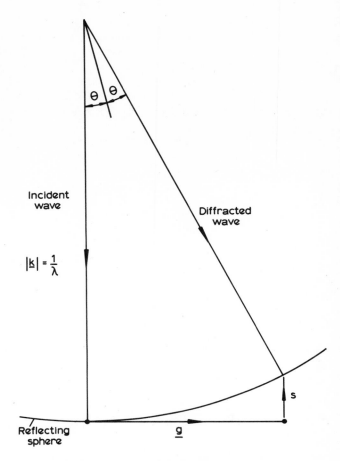

2 Ewald sphere construction for determining the direction of a diffracted wave

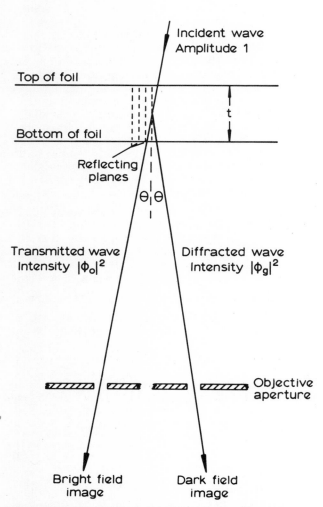

1 Transmission and diffraction of an electron wave incident on a foil. Bragg angle Θ is about 10^{-2} radians

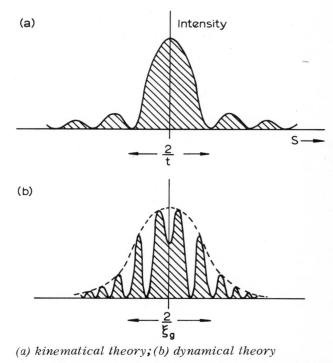

(a) kinematical theory; (b) dynamical theory

3 Intensity distribution (schematic) around a reciprocal lattice point

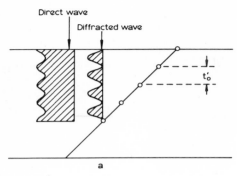

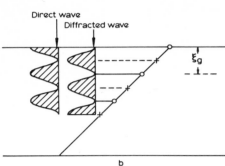

(a) kinematical region where the diffracted beam is weak; (b) dynamical region where s = 0

4 Intensity oscillations (schematic) of direct and diffracted waves in a crystal

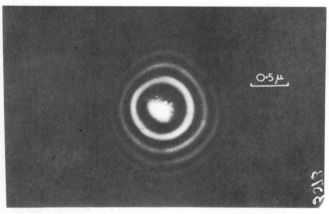

5a Thickness extinction contours in Al; roughly parallel pairs arise from reflections from opposite sides of Bragg planes

5b Bend contours in Al due to 220 and 311 type reflections

to atomic displacements. The contrast arising is therefore like phase contrast, the phase change being due to atomic displacements.

For a dislocation R_n may be estimated by continuum elasticity theory. The images of screw dislocations calculated by kinematical theory are then as shown in Fig. 6. The parameter on each curve is an integer $n(=g.b)$, where b is the Burgers vector of the dislocation. Usually n has the value 0, ±1, ±2. We note that the image peak lies to one side of the dislocation core, and that the abscissa is the product of s and x (distance on the image). As a result images are expected to become narrower if s is made larger. Use of this fact is made in the so-called 'weak-beam technique' for observing defects at higher resolution.

Near the Bragg position the kinematical theory breaks down. In this region the so-called 'dynamical theory' takes over. The dynamical theory describes correctly the 'dynamical equilibrium' between the various waves in the crystal, and takes account of the depletion and augmentation of various waves by Bragg scattering. For the two-beam theory, the incident and diffracted wave amplitudes $\phi_0(z)$ and $\phi g(z)$ are described by the following coupled differential equations:

$$\frac{d\phi_0}{dz} = i\frac{\pi}{\xi_0}\phi_0 + i\frac{\pi}{\xi_g}\phi_g \exp(2\pi isz + 2\pi ig.R)$$

$$\frac{d\phi_g}{dz} = i\frac{\pi}{\xi_g}\phi_0 \exp(-2\pi isz - 2\pi ig.R) + i\frac{\pi}{\xi_0}\phi_g$$

$$\cdots\cdots(4)$$

The quantity ξ_g is the extinction distance (Fig. 4b). ξ_0 is related to the refractive index. For a perfect crystal, $R = 0$, and equation (4) may be solved analytically to good approximation to give the well known

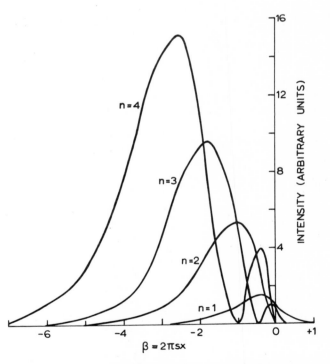

6 Kinematical intensity profiles of dark-field images of a screw dislocation for various values of the parameter n = g.b. Core of the dislocation is at $\beta = x = 0$

equations of two-beam dynamical theory. If $R(z)$ is known for a column of crystal at distance x from a defect such as a dislocation, the equations may be integrated by digital computer, with the boundary conditions at the top surface $\phi_0(0) = 1$, $\phi_g(0) = 0$. The problem is to find $\phi_0(t)$ and $\phi_g(t)$ at the bottom of the foil. The intensity of bright-field and dark-field images is then given by $I_0 = |\phi_0(t)|^2$ and $I_g = |\phi_g(t)|^2$. The calculation is repeated at different distances x from the dislocation and $I_0(x)$ and $I_g(x)$ give the profiles of bright-field and dark-field images of the dislocation (Fig. 10). The calculation may be repeated for the dislocation at different depths y in the foil to simulate an inclined dislocation. It can be seen that the number of parameters can be quite large. We may write

$$I_g = \text{function of } (t, x, y, \xi_g, \xi'_g, s_g, \text{strain field})\ldots(5)$$

The strain field itself may incorporate several parameters particularly if anisotropic elasticity is involved. We have not mentioned the additional parameter ξ'_g which takes account of absorption. Equation (4) may be generalized to describe the n-beam situation, in which case there are n values of g, ξ_g and ξ'_g.

To sum up, the interpretation of a particular image contrast effect from a defect involves the following procedures:

(i) postulation of a model of the defect
(ii) calculation of the strain field and lattice displacements
(iii) calculation of the image contrast from equation (4)
(iv) display of the calculated results
(v) comparison with experimental observations
(vi) further modification of the model and recalculation.

It is clear that a systematic study will necessarily need to employ a fast digital computer for steps (ii), (iii), and (iv) to investigate all the parameters which may be varied. Often to obtain further confirmation of a proposed model it will be necessary to obtain more experimental images with different Bragg reflections, which are then compared with more calculations. The further images required are often suggested by the results of preceding calculations.

For a review of the theory outlined here the reader should consult the reference.[1]

EXAMPLES OF APPLICATIONS OF THE THEORY

The simplest example is the case of a planar defect such as a stacking fault shown schematically in Fig. 7. A perfect crystal has been sheared on a plane at depth t_1 by a vector R which is not a lattice translation vector. In the f.c.c. structure faults usually occur on $\{111\}$ planes and the phase angle α ($=2\pi g.R$) takes values 0, $\pm 2\pi/3$ (modulo 2π) depending on the order g of the reflection. The intensities $|\phi_0|^2$ and $|\phi_g|^2$ in Fig. 7 are functions of t, t_1, α, s. For an inclined fault t_1 varies across the fault, and the intensity across the fault is in the form of fringes as illustrated in Fig. 7. Fig. 8 shows an example of computed fringes, while Figs. 9a and 9b show experimental images of faults in Cu + 7% Al alloy. Both the computations and the experimental images demonstrate the fact that the bright-field image is

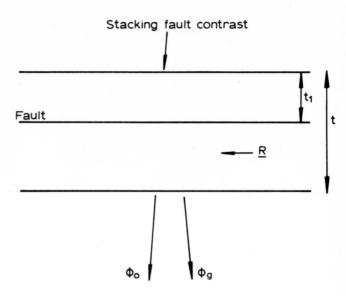

Phase angle $\alpha = 2\pi \, \underline{g}.\underline{R}$

ϕ_0 ϕ_g – functions of (t, t_1, w, α)

Eg. inclined fault

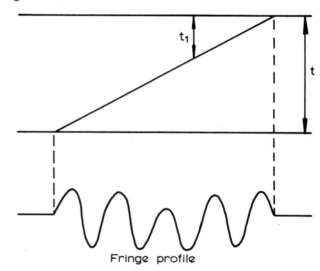

7 Schematic diagram of a stacking fault in a thin foil

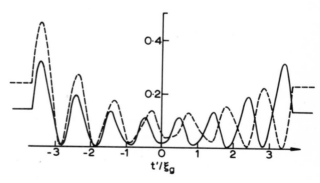

$\alpha = {}^{+}2\pi/3; \; t/\xi_g = 7\cdot25; \; \xi'_0 = \xi'_g; \; \xi_g/\xi'_g = 0\cdot075;$
$\xi_g s = -0\cdot2.$

8 Profiles of stacking fault fringes computed to match Fig. 9. Full curve is bright-field image. Broken curve is dark-field image

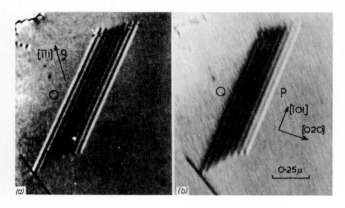

9 *a* bright-field and *b* dark-field images of a stacking fault in f.c.c. Cu + 7 wt.% Al alloy. Foil normal is [101]

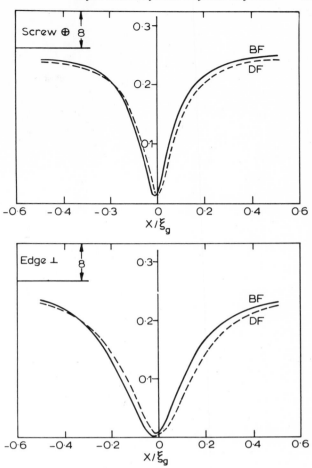

$g.b = 1, s = 0$

10 Computed image-profiles of screw- and edge-dislocations in the centre of a foil of thickness $t = 8\xi_g$

symmetric about the centre (i.e. edge fringes are similar), while the dark-field image is asymmetric. This fact can be used to determine the sense of inclination of the fault, e.g. P in Figs 9a, 9b, is the edge where the electron beam enters the foil. The sense of the edge fringe in the bright-field image depends on the sign of α. These facts enable one to determine the nature of the stacking fault in the f.c.c. structure, i.e. to determine whether the fault is of the intrinsic or extrinsic variety. The appearance of the fringes at faults, particularly their branching behaviour as the thickness increases in a wedge crystal, depends in detail on the value of α. Another important value of α encountered is the case $\alpha = \pi$, which has been found for certain organic structures such as Al N[2]. A careful study of the fringes enables one to differentiate between $2\pi/3$ and π faults[2] (for review see reference[3]). One point that should be noted is that faults are not always visible.[4,5] If the phase angle α is a multiple of 2π, as it may well be for certain reflections, the fault will be invisible. Vanishing criteria can therefore be used as an aid in determining the fault vector.

The problem of dislocation image contrast can be solved by integrating the equations (4), numerically by digital computer, after substitution of suitable displacements R. Numerical integration methods such as the Runge-Kutta or Adams processes are usually employed. As in the kinematical theory, image profiles are characterized by an integer $n = g.b$, which usually takes values of 0, ±1, ±2. Figure 10 shows computed image profiles of screw and edge dislocations for the case $n = 1$. The image width of screw dislocations is about $\xi_g/3$, and is slightly wider for an edge dislocation. For the case $n = g.b = 0$, a screw dislocation will be invisible, i.e. the displacements due to the dislocation lie in the Bragg reflecting planes and therefore do not affect the intensity of the reflection. A similar invisibility criterion holds approximately for an edge or mixed dislocation. The vanishing of a dislocation in different reflections can be used as a method of determining the direction of the Burgers vector b of the dislocation. The magnitude and sense of the Burgers vector can be found by detailed comparison of the experimental images with computed images (as will be discussed). We note in passing the problem of determination of the nature of small dislocation loops, i.e. to determine whether they are interstitial or vacancy in character. This may be done if the sense of the Burgers vector on a suit-

able convention is determined. One can make use of the fact that if the foil is tilted away from the Bragg reflecting position the image peak lies to one side of the dislocation core, so that the image of a loop lies entirely inside or entirely outside the core of a closed loop (see Fig. 14). Evidently the answer will depend on several factors, such as the direction of g, the inclination of the loop plane, and the sign of s. Careful experiments enable the loop sense to be determined.[6] We also note the similar problem of determining the nature of small, spherically symmetric inclusions,[7,8] and the nature of radiation damage clusters in irradiated materials.[9,10,11]

COMPUTER SIMULATION OF IMAGES

The study and explanation of special contrast effects at inclined dislocations, such as at A and B in Fig. 11 requires the computation and adjacent display of many computed image profiles for the dislocation at various depths in the foil as in Fig. 12. Even with such a display it is difficult to form a mental impression of how the image should look. In recent years a very useful method of computer display of the cal-

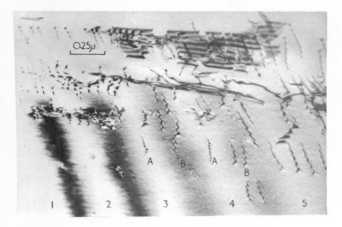

1, 2, 3 ... regions of thickness $t = \xi_g, 2\xi_g, 3\xi_g ...$
Note the dotted and zig-zag appearance of dislocation
images at A and B

11 Transmission electron micrograph of dislocations
and faulted regions in a foil of Cu + 7 wt.% Al alloy

culated image has been developed, originally by Head[12]
and later by others.[13,14,15,16,17] In Head's original
technique (Fig. 13) the image is displayed by arranging
for the computer line printer to produce a 'grey scale'
by overprinting type symbols. Thus 'dots' of varying
density corresponding to the ordinate on the profile
curves may be used to construct a two-dimensional
display of the image. Figure 13a—d shows examples
(due to Head[12]) of experimental images of screw
dislocations in β-brass, side by side with computer-
simulated images for the various reflections indicated.
A comparison of computed and experimental images
enables one to decide that the Burgers vector is $[\bar{1}11]$
rather than $\frac{1}{2}[\bar{1}11]$. The technique has been improved
by Bullough, Maher, and Perrin,[15] who have used a
cathode ray tube curve-plotter to produce simulated
images of dislocation loops which are almost in-
distinguishable in quality from experimental images.
The number of 'dots' forming the picture can be
increased while the steps in the 'grey scale' can be
made finer. A similar method has been used by Foreman and Hudson[16] to display computed images of
bend extinction contours in foils of varying thickness
and at different accelerating voltages.

Figure 14 shows a composite picture of which the
upper pair was produced by computer simulation[15]
and the lower pair shows experimental micrographs
of dislocation loops in aluminium irradiated with
α-particles.[16] In both these pictures the loops are
observed with s positive (the reciprocal lattice point
inside the Ewald sphere in Fig. 2), but the sign of g is
changed by tilting between members of each pair.
The pictures demonstrate clearly the change in size
of the loops on tilting from which the sense of the
loops (vacancy or interstitial) can be determined.[6]
Figure 15 is due to Karnthaler, Hazzledine, and
Spring,[17] and shows computer-simulated images
(outlined) of dissociated edge- and screw-dislocations
for various reflections indicated displayed on actual
micrographs of such dislocations in Cu + 10 wt %
Al alloy. The computed images illustrate how well
theory accounts for details of dislocation image contrast, which are only interpreted with difficulty using
displays like Fig. 12.

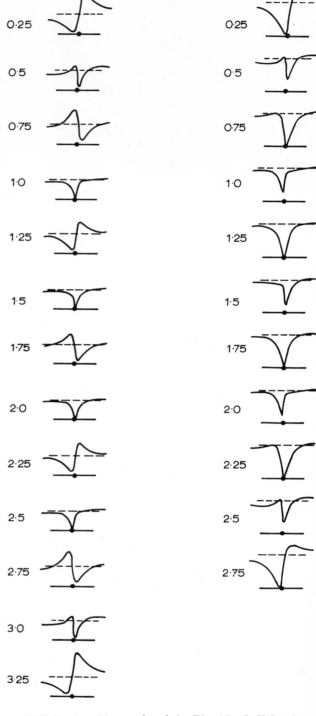

R.H.S:- $t = 3\xi_g$ as for A in Fig. 11. L.H.S:- $t = 3.5\xi_g$ as for B in Fig. 11

12 Computed image-profiles for a screw dislocation
depths given in extinction distance units on each
profile

HIGH-RESOLUTION IMAGING TECHNIQUES

The electron microscope can also be used as a tool
for investigating lattice defects at high resolution.
At the highest resolution an attempt is made to
resolve the lattice planes of the structure by including
two or more diffracted beams in the image as in

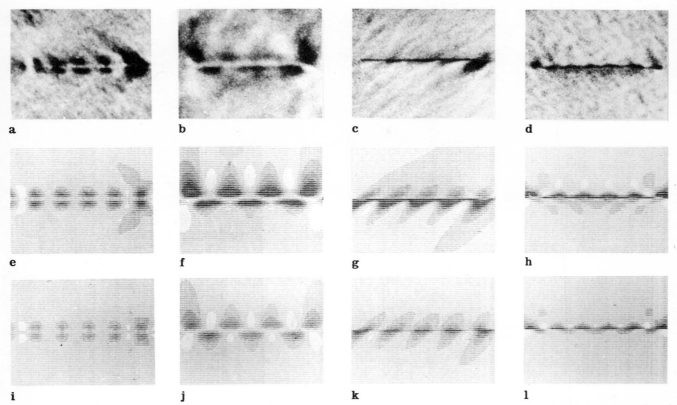

(a)-(d) experimental images for the reflections indicated; (e)-(h) computer-simulated images for $\underline{b} = [\bar{1}11]$; (i)-(1) computer-simulated images for $\underline{b} = \frac{1}{2}[111]$

13 Comparison of experimental and computer-simulated images in β-brass

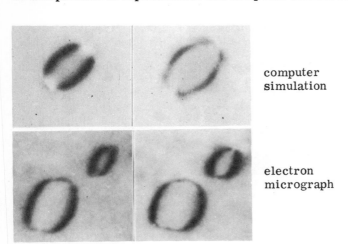

computer
simulation

electron
micrograph

14 Experimental and computer-simulated images of dislocation loops[6,15]

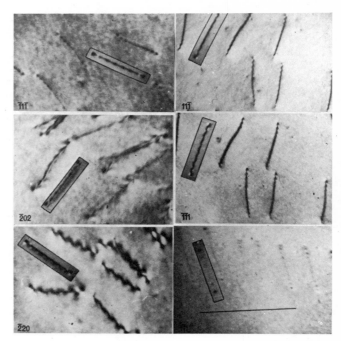

15 Computer-simulated images (outlined)

Abbe's theory of resolution of a grating in physical optics. An example is shown in Fig. 16 (due to Cockayne, Parsons and Hoelke[18]) where the {111} planes in germanium of 3·27 Å spacing are resolved. Strictly speaking these should be called 'lattice fringes' rather than 'images of the atom planes'. The reason for this distinction is that there is not always a one-to-one correspondence between the actual atom planes in the structure and the lattice fringes formed by diffraction from these planes. In particular, terminating fringes as in Fig. 16 cannot always be identified in a simple way with terminating half-planes in the crystal structure at edge dislocations.[18] Figure

16 illustrates this point where a change of the diffraction conditions from (a) to (b) changes the number of half-fringes visible at an inclined screw dislocation from one to three. Evidently, care must be taken in interpreting details of lattice fringes in terms of strains and dissociations.[19,20]

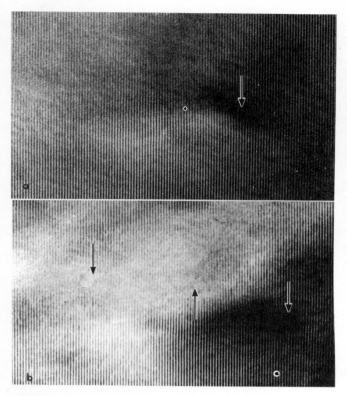

16 {111} lattice fringes of spacing 3·27 Å in germanium

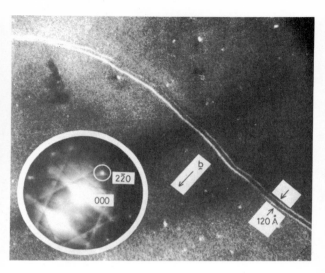

17 Weak beam image of a dissociated edge dislocation

Aiming at a somewhat lower level of resolution, a weak-beam imaging technique has been developed by Cockayne, Ray and Whelan,[21] which has proved very useful in investigating geometrical aspects of dislocation dissociations. This technique employs the result, mentioned in the outline of theoretical aspects, and illustrated in Fig. 6, that under kinematical conditions, the width of the dark-field image of a dislocation becomes very narrow as the deviation from the Bragg condition increases. In the dark-field image the peak intensity is also many times background intensity, so that the images are clearly visible. Thus it becomes possible to resolve closely spaced partial dislocations resulting from dissociation which would not be resolved using strong-beam images owing to the relatively large image width ($\sim \xi_g/3$). The image widths of weak-beam images of dislocations may be as low as 10 to 15 Å. Figure 17 shows an example in Cu + 10% Al alloy, where the dissociation into partials of a dislocation close to edge orientation is visible. From measurements of the separation of the partials it is possible to estimate the stacking fault energy in this alloy as a function of alloy composition.[22] Further applications of the method to the study of dissociated dislocations in silicon[23] and to the study of dislocations in B2 and DO_3 superlattices in the Fe_3Al system[24,25] have been made. High-resolution studies of this type are proving very useful in the study of the geometry of dissociation from which important quantitative information on stacking fault energies can be derived.

REFERENCES

1 **P. B. Hirsch** *et al.*: 'Electron Microscopy of Thin Crystals', 1st edition, 1965, London, Butterworths
2 **C. M. Drum and M. J. Whelan:** *Phil. Mag.,* **11,** 205
3 **S. Amelinckx** *et al.*: (Editors): 'Modern Diffraction and Imaging Techniques in Material Science', 35-98; 1970, Amsterdam, North-Holland Publishing Company
4 **M. J. Whelan and P. B. Hirsch:** *Phil. Mag.,* **2,** 1121
5 **M. J. Whelan and P. B. Hirsch:** *Phil. Mag.,* **2,** 1303
6 **D. J. Mazey** *et al.*: *Phil. Mag.,* **7,** 1861
7 **M. F. Ashby and L. M. Brown:** *Phil. Mag.,* **8,** 1083
8 **M. F. Ashby and L. M. Brown:** *Phil. Mag.,* **8,** 1649
9 **M. Rühle** *et al.*: *Phys. Status Solidi,* **11,** 819
10 **K. G. McIntyre:** *Phil. Mag.,* **15,** 205
11 **M. M. Wilson:** *Radiation Effects,* **1,** 207
12 **A. K. Head:** *Austral. J. Phys.,* **20,** 557
13 **A. K. Head** *et al.*: *Phys. Status Solidi,* **20,** 505
14 **P. Humble:** *Austral. J. Phys.,* **21,** 325
15 **R. Bullough** *et al.*: *Phys. Status Solidi,* (in press); AERE rep. R6348, 1970, and TP406, 1970
16 **A. J. E. Foreman and B. Hudson:** Proc. 7th Int. Cong. Electron Microscopy, 77-78, 1970, Paris, Société Française de Microscopie Electronique
17 **H.-P. Karnthaler** *et al.*: *Phil. Mag.,* to be published
18 **D. J. Cockayne** *et al.*: *Phil. Mag.,* to be published
19 **J. R. Parsons and C. W. Hoelke:** J. Appl. Phys., **40,** 866
20 **V. A. Phillips and J. A. Hugo:** *Acta. Met.,* **18,** 123
21 **D. J. H. Cockayne** *et al.*: *Phil. Mag.,* **20,** 1265
22 **I. L. F. Ray:** D. Phil. Dissert., 1971, University of Oxford
23 **I. L. F. Ray and D. J. H. Cockayne:** *Phil. Mag.,* **22,** 853
24 **I. L. F. Ray** *et al.*: *Phil. Mag.,* **21,** 1027
25 **R. C. Crawford:** D. Phil. Dissert., 1971, University of Oxford

USE OF COMPUTERS IN ELECTRON PROBE MICROANALYSIS

P. Duncumb

Over the past decade, the advent of easy-to-learn programming languages such as 'Fortran' has made the computer an invaluable tool to users in many different disciplines. With its aid, the physicist has been able to study in some detail the processes that take place in a metal which is bombarded with an electron beam, and in consequence, the metallurgist now has computer programs to give a quick and accurate interpretation of quantitative measurements made with the electron probe microanalyser. The purpose of the paper is to show, from the users point of view, what is now possible in electron probe analysis with the aid of the computer. To do this, some of the physical models which have been used to simulate the complex electron-target interactions, the way in which they may be evaluated, and some of the attributes of a successful computer program for representing a selected correction procedure to the microprobe user are considered. Brief mention is also made of the use of the computer on-line to the microprobe instrument, for data acquisition and programmed control.

The author is with the research laboratories of Tube Investments Ltd

620.187.22:519.5

COMPUTER MODELS OF ELECTRON-TARGET INTERACTION

The central problem in describing the generation of X-rays mathematically lies in the manner in which electron scattering within the target is taken into account.[1]

As an electron penetrates the target it is slowed down almost continuously by a large number of inelastic collisions with orbital electrons, and is scattered elastically by a similarly large number of nuclear (Rutherford) interactions. Only a few of the Rutherford scattering acts cause the electron to be deflected through large angles, but the overall effect of a succession of small-angle events is to produce a random diffusion in the target, in which electrons, towards the end of their trajectories, move in random directions unrelated to that of the incident beam. This progressive onset of diffusion below the surface is one of the main factors controlling the distribution of X-ray production with depth within the target, and hence the overall absorption which the X-rays undergo in the target as they emerge in the direction of the spectrometers. The scattering process can also result in some electrons being backscattered out of the target surface, carrying with them energy that would otherwise have been used in producing X-rays.

The first attempt that was made to simulate this scattering process in the computer was by Archard and Mulvey,[2] who proposed the 'single scattering' model. Electrons are assumed to pass into the target undeflected, though losing energy continuously, until they reach a depth equivalent to that of complete diffusion. At this point they are assumed to be scattered in all directions with equal probability until they come to rest within the target, or leave the target surface. This is shown schematically in Fig. 1, although because of the spherical symmetry, the paths shown do not represent equal numbers of electrons. Given the electron retardation as a function of energy, the length of the trajectory may be calculated and also the electron energy and X-ray production at each point along it. The overall distribution of X-ray production as a function of depth z below the surface may thus be obtained by summing the contributions from each trajectory within a layer $z \rightarrow z + dz$ beneath the surface. This approach was tested with moderate success by Archard and Mulvey, and more recently by Wolf and Macres.[3] The model is attractive in that it can take account easily of non-normal electron beam incidence, but suffers from the obvious gross approximation of representing all the scattering by one event.

A more realistic approach to the problem was later made by Bishop,[4] who employed a Monte Carlo technique to simulate the succession of scattering acts in each trajectory. The trajectory length was divided into a number of steps, 25 in his work, and the angle of scattering at each step computed from a random number, which was weighted according to the known scattering laws for elastic interactions. In this way, a large number of trajectories could be generated, each different from all the others, from which the backscattering and X-ray production could be summed as before. In order to smooth out the random nature of this calculation a large number of trajectories (up to 5 000) were normally calculated, so that the resulting corrections for backscatter, etc., represented a true mean value to within 1%. This approach is ideally suited to a large computer, and has been converted into a routine correction procedure by Pascal,[5] using a CDC 6600.

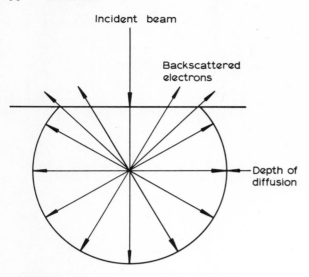

1 Archard-Mulvey model of electron scattering

It is, however, possible to obtain useful information from a small machine, such as the IBM 1130, by simplifying the scattering formula to enable it to be computed at each step, instead of storing a large array of previously computed angles of scatter, as is done in the above work. An example of some results obtained by Duncumb and Curgenven (unpublished) on an IBM 1130 is given in Fig. 2, which shows the paths of 100 electrons in a copper target, with the incident 20kV beam inclined at 45° to the target surface. This diagram was plotted directly on the computer, and may be produced in two colours representing the portion of each trajectory above and below the critical excitation potential, i.e. distinguishing the part from which X-rays are produced. The onset of diffusion is clearly visible, and the diagram also indicates that the number, direction, and energy of the backscattered electrons may be obtained, as well as those at various depths within the target.

Like the single scattering model, the Monte Carlo technique is particularly valuable for unusual target geometries, not only for inclined beam incidence, but also for varying boundary conditions in the target itself. In a thin film target, for example, the degree of scattering taking place within a known thickness of target may be obtained, as is illustrated in Fig. 3. In the thinnest target (0·25 μm), very little lateral scattering takes place: the electrons pass into the target and out the other side with little loss of energy.

In the 1·3 μm sample, the electrons are still transmitted, but a considerable degree of backscatter is taking place: almost equivalent to that which arises in a thick target. Thus the Monte Carlo approach can be usefully applied to the study of conditions such as this, but is unlikely to be widely used for routine correction of results, because of the very large amount of computing required to give statistically accurate answers.

Mention may be made at this point of the transport equation program (TEP) developed by Brown, Wittry, and Kyser,[6] in which the electrons are treated as a flux rather than as individual particles, and the onset of diffusion is described by a modified Boltzmann diffusion equation. Good agreement has been obtained with experimental results, but it is too soon to say whether the somewhat lengthy computation (though not as long as Monte Carlo) is justified by the accuracy obtained.

The most widely used model for representing electron-target interactions is an extension of that proposed by Philibert,[7] now known as the ZAF model,[1] since it takes separate account of target atomic number (Z), X-ray absorption (A), and fluorescence (F). This employs a simplified description of the electron scattering process but lends itself to empirical adjustment to match experimental results, and may be represented in the form of a computer program for the microprobe user which is both fast and economical in computer storage. The ZAF approach assumes the same laws of electron retardation as the Monte Carlo, but uses experimental values for electron backscatter, and describes the electron diffusion process as a progressive increase in mean angle of the elec-

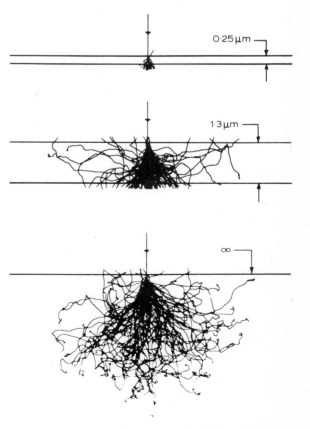

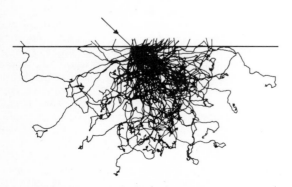

2 Copper target at 20kV. 45° beam incidence

3 Aluminium target at 20kV. Varying thickness

tron flux to the direction of the incident beam. The
X-ray production in a given layer in the target is
then derived from the number of electrons arriving
at this layer and the angle at which they cross it.
Because the electrons are not treated individually
the amount of computation involved is less, but the
simplifications introduced require that the method
be thoroughly tested in practice, with a final empirical
adjustment of the least known parameters.

This optimization process is best carried out from
the error histogram obtained from the corrected
analyses of a large number of samples of known composition, chosen from as wide a range of atomic number, operating voltages, and as many different laboratories, as possible. In this way the random element
in imperfect experimental measurements may be
separated from systematic errors in the simplified
theory. The final adjustment of each key parameter
in the theory is made to minimize the rms error
in the histogram of a given class of samples, selected
to be sensitive to the parameter under consideration.
For example, in the absorption correction, the rate at
which the number of electrons available for X-ray
production falls off with increasing depth is chiefly
governed by the effective Lenard coefficient σ. The
precise dependence of σ on incident beam energy is
therefore best determined from a series of samples
having a high absorption correction, such as occurs
in silicate minerals; these samples have the added
advantage of being homogeneous and of accurately
determinable composition. Similarly, some inter-
metallic oxide and carbide samples containing elements of widely differing atomic number may be used
for adjusting the ionisation potential J appearing in
the atomic number correction. The choice of both
σ and J has been discussed by Duncumb, Mason, and
da Casa, [8] and, as an example, Fig. 4 shows the error
histogram of an uncorrected set of silicate measurements, and of the same set corrected, using two forms
of the σ parameter, proposed by Duncumb and by
Heinrich. In this trial, it emerged that Heinrich's
form (chosen on the basis of more recent experiments) was superior.

An optimization process of this type is ideally suited
to interactive computer operation. The experimental
data are stored on disk in the computer and processed over and over again as the appropriate parameter is adjusted iteratively to the optimum value.
After each trial the operator can evaluate the result
either as an rms error or by plotting the histogram
on the computer plotter, and can decide on the basis
of the result what class of samples and what value
of parameter to select next. This decision process
is impossible to program into the computer, since
the results often throw up unlooked-for relationships
which can only be interpreted by an experienced
microprobe operator. The whole process is therefore
greatly speeded if this person can control the running
of the program on the computer himself.

One other correction procedure may be referred to
at this stage, although it is not based on a physical
model of the X-ray production process. This is the
'alpha method', which has been presented in several
forms, notably by Ziebold and Ogilvie, [9] and which is
widely used in X-ray fluorescence analysis. The
relation between the measured (uncorrected) concentration k_A of element A, and true concentration c_A, is

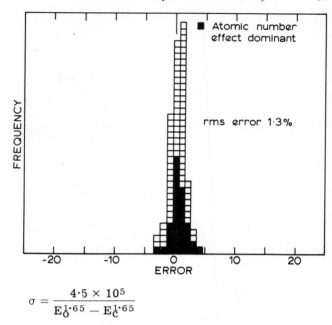

$$\sigma = \frac{4 \cdot 5 \times 10^5}{E_0^{1 \cdot 65} - E_C^{1 \cdot 65}}$$

4a Corrected 121 silicate analyses using Heinrich σ

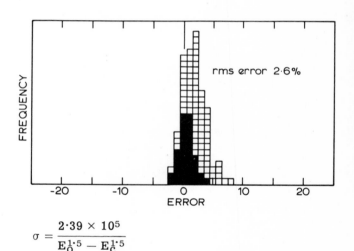

$$\sigma = \frac{2 \cdot 39 \times 10^5}{E_0^{1 \cdot 5} - E_C^{1 \cdot 5}}$$

**4b Corrected 121 silicate analyses using Duncumb/
Shields σ**

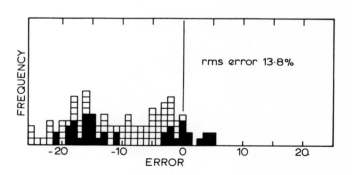

4c Uncorrected 121 silicate analyses

derived by summing the effect of a number of inter-element α parameters, defined by the expression

$$k_A = c_A / \sum_i c_i \alpha_{Ai}$$

A unique α parameter is determined, either empirically or theoretically, for each pair of elements present in the sample, so that in the case of a binary system, for example, the entire k to c calibration curve for element A is defined by a single parameter α_{AB}:

$$k_A = \frac{c_A}{c_A + \alpha_{AB}(1 - c_A)}$$

This form of equation gives rise to a set of curves of the type illustrated in Fig. 5, showing the calibration curves for each element of a binary system at a number of different beam energies. These curves were actually calculated by the ZAF method, and plotted by computer, but each curve can be closely characterized by a single α value. In this example, the measured concentration k_A is always less than the true concentration c_A, so that α is always greater than unity, but this is not the case when fluorescence is the dominant correction, causing an enhancement of the measured value.

For a complex sample containing several elements, a matrix of α coefficients is required as a starting point, but the solution of a set of $c_A, c_B, c_C \ldots$ values from experimental $k_A, k_B, k_C \ldots$ values is then simply obtained by a matrix inversion. This is a process which can be carried out on a small computer of limited storage capacity ($<4k$ words for up to say, 10 elements), and which may therefore be attractive to a user having a 'mini-computer', perhaps dedicated to the microprobe for control purposes. Some loss of accuracy is to be expected, but in the trial on the silicate analyses illustrated in Fig. 4, for example,

the rms error is increased only from 1·3% to 1·5% The conclusion that accuracy is in general only little affected, however, has yet to be checked more fully. In practice most users at present will be correcting data on a medium or large machine, and the next section is concerned with the type of program most suited to this purpose.

COMPUTER PROGRAMS FOR ROUTINE ANALYSIS

A key point about a successful program for routine correction of results is that it must be easy to use by non-computer personnel. If the type of program referred to above is analogous to a laboratory 'lash-up' (which may work perfectly, but usually only for the person who wrote it), the program for routine analysis must be equivalent to a fully proven product, foolproof in every respect. Clear documentation of the program is important, and, whilst a number of user options may be valuable, these must in no way complicate the normal way of running the program. It is sensible as far as possible to use the computer to store fixed constants, such as mass absorption coefficients, atomic weights, critical excitation potentials, etc., so that the user has to input the very minimum of data.

By far the most popular type of program at the present time is that based on the ZAF model. Nearly 40 such programs have been written, and these have been classified and compared in detail by Beaman and Isasi.[10] Whilst it is possible to some extent to make recommendations for desirable features in a correction program, these authors are at pains to emphasise that the type of work to be carried out and type of computer available largely account for the existence of so many. Nevertheless, four such programs are commended in the overall comparison: Mason, Frost and Reed,[11] Duncumb and Jones,[12] Shaw,[13] and Colby.[14]

To illustrate the minimal effort which is required on the part of the microprobe operator in using such a program, the data cards which must be provided for the program by Duncumb and Jones, in a typical analysis of a stainless steel, are shown below:

P. DUNCUMB AND E.M. JONES, PHYSICS DEPT.
EMMA-3
FEK1 CRK1 NIK1
30.0   KV
70.4 21.2 6.6   MK

Five cards only are required, the first two simply identifying the operator and instrument respectively. Card 3, the system card, specifies the radiations measured and hence the elements present (maximum 8), the last named element being assumed to be calculated by difference if the radiation is left unspecified. The numbers following the radiation indicate the standard used, thus FeK1 signifies FeKα radiation analysed against standard 1 (pure iron). Complex standards are referenced by a higher number, and the compositions are previously stored in a disk file in the computer. Mass absorption coefficients and element constants are stored in further files and are looked up without any further instruction from the operator. Card 4 gives the kilovoltage at which the analysis was carried out (which may be

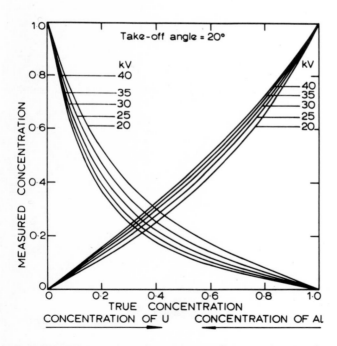

5 Calibration curves for measured concentration k_A against true concentration c_A

different for each element), and Card 5 the measured concentration (in percentage) for each measured element. Some programs enable direct count rate figures for peak and background to be entered, from which the measured concentration may be calculated. This facility is not available on this program, but a number of other options, such as expressing the answer in terms of percent of equivalent oxide, are available. These option cards may be entered after the identification cards 1 and 2, if required.

The output from the program in this example is then as shown below:

P. DUNCUMB AND E. M. JONES, PHYSICS DEPT.
EMMA-3
SYSTEM FEK 1 CRK 1 NIK 1
ANALYSIS NO. 1

	KV	MEASURED	TRUE
FE	30·0	70·400	73·937
CR	30·0	21·200	17·975
NI	30·0	6·600	7·988
		TOTAL	99·900

This gives the calculated concentration for each element; the individual correction factors may be printed if desired. Provision is made to enter a batch of analyses in one run, and the output tabulated page by page for easy reading. This particular program is intended for use on an IBM 1130 computer having 8k core storage plus a single disk backing store. The version by Mason, Frost, and Reed is based on the same theory but makes use of the much larger IBM 7094, and offers many more optional facilities to the operator.

USE OF COMPUTERS ON-LINE

The recent advent of cheap 'mini-computers', primarily intended for control and data acquisition, is now making an impact on the electron probe microanalyser field. Microprobe instruments are expensive, as are operators, and the comparatively low cost of these devices makes use of the much larger of automation very attractive.

It is too soon to say how far this will be carried, but already at least one commercial system, with interface, is available with software for attachment to the microprobe. Control of the spectrometers and specimen stages may be carried out by stepper motors, which are easily driven from digital logic equipment, if somewhat slow in operation. Good mechanical repeatability is, of course, presupposed in the instrument, but backlash can be largely removed by programming the drive motors to approach the desired wavelength or specimen position always from the same direction. Analog data, such as probe current, is converted into digital form by A-D converters, whereas scaler measurements of X-ray intensity are already in a suitable form for entry into the computer. Instructions from the operator, for example the elements to be analysed at a number of preselected points, may be entered by Teletype or prepared previously on paper tape, and the same console used to print out results. Thus the computer performs the function both of data acquisition and control, and offers the possibility of closed-loop operation, e.g. modifying the next control instruction in the light of

data just acquired. A statistical test on a measured count may, for example, be made before proceeding to the next measurement. The computer may also be connected to a larger computer for processing the data and correcting the results.

One example of this kind of application has been demonstrated by Colby (unpublished). The problem here was to determine the component peak positions of the SiKβ emission in various silicon-containing materials, in order to relate this information to chemical valency. The spectrometer was programmed to scan over the SiKβ emission in 200 steps of 10-second counting periods, the results of each count being sent back by data link from the on-line computer to a large machine. This calculated the standard deviation of each measurement, applied a smoothing procedure to the whole spectrum, and by multiple differentiation established the component peak positions. The spectrum was finally sent back to a plotter located near the microprobe, with peak positions clearly marked. In the example given in Fig. 6, the two main components were widely separated; in other cases, overlap can cause an apparent peak shift, which may be readily detected in the differentiation process referred to above.

Clearly much remains to be learned about the optimum modes of operation of such a powerful combination. At one extreme it is easy to see how repetitive measurements, such as point-by-point analyses of known elements, can be facilitated. At the other, there still remains much exploratory work on unknown specimens in which it is difficult to see how human interpretation during the investigation can be avoided. It seems reasonable to suppose, however, that the logic of probe operation will eventually be formalized sufficiently to write a control program with some considerable degree of flexibility. One might specify, for example, that all particles larger than 10 μm in a steel sample were to be analysed for Mn, Si, and Al, the results of each analysis corrected, and if the total, expressed as oxides, lies between 98% and 102%, then the particle is to be counted as an alumino-silicate. If this is not the case, however, a search could be

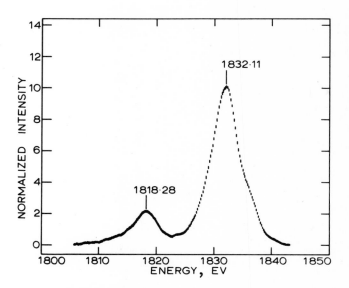

6 SiKβ emission spectra from fused quartz

made for Ca, to assess the contribution from slag, or failing this, for K or Na to detect the presence of mould additive. Although the final result may take longer to obtain than with a human operator, it may require considerably less human effort and be run overnight to make best use of the instrument. Several laboratories are at present studying the application of the computer to electron probe microanalysers, and the next few years will undoubtedly show progress in this field.

CONCLUSIONS

Several types of computer applications to electron probe microanalysis can now be recognized. First, there is the mathematical representation of the physical model used to describe electron-target interactions taking place. Secondly, there is the testing and optimization of these models against experimental data. Thirdly, the selected model must be put in the form of a proven correction program, easy-to-use by the microprobe operator with little computer experience. Finally, there is the relatively new development of data acquisition and control with computers on-line to the electron probe microanalyser itself. Many other computer-related topics could be included, such as electron-optical lens design, but these main classes serve to categorize computer applications from the stand-point of the user. From his point of view it is encouraging that, although some degree of computer awareness is becoming essential, much can be achieved with little or no special ability in computer programming, leaving him free to concentrate his attention on the problem itself.

ACKNOWLEDGMENTS

The X-ray spectrum illustrating the use of computer control was kindly provided by Dr J. W. Colby, of Bell Telephone Laboratories; and the electron trajectory plots by Mrs Lesley Curgenven of these Laboratories. The author is indebted to the Chairman of Tube Investments for permission to publish this paper.

REFERENCES

1   J. Philibert: Proc. 5th Int. Cong. X-ray Optics and Microanalysis, Berlin, Springer, 1969, 114
2   G. D. Archard and T. Mulvey: *Brit. J. App. Phys.*, 14, 626
3   R. C. Wolf and V. G. Macres: Adv. in Electronics and Electron Physics, Academic Press 1969, Suppl. 6, 73
4   H. E. Bishop: *Brit. J. App. Phys.*, 18, 703
5   B. Pascal: Ref. 1, 135
6   D. B. Brown *et al.*: *Brit. J. App. Phys.*, 40, 1627
7   J. Philibert: Proc. 3rd Int. Cong. X-ray Optics and Microanalysis, Academic Press, 1963, 379
8   P. Duncumb, *et al.*: Ref. 1, 146
9   T. O. Ziebold, and R. E. Ogilvie: *Anal. Chem.*, 36 322
10 D. R. Beaman and J. A. Isasi: *Anal. Chem.*, 42, 1540
11 P. K. Mason *et al.*: NPL Rep. No. IMS 2, 1969
12 P. Duncumb and E. M. Jones: Tube Investments Technical Rep. 260
13 J. L. Shaw: AERE Rep. R6071
14 J. Colby: *Adv. in X-ray Analysis*, 11, 287

USE OF COMPUTERS IN CRYSTALLOGRAPHY

W. Johnson

Computers have been applied to many areas of crystallography and basic program problems have been solved, although the regular application is still confined to a comparatively small number of laboratories. The application of computers to control equipment in crystallography is still not widely applied. In most cases the computer performs the function of digesting and reducing the amount of information, except where visual display is intended. Where large amounts of information are calculated by the computer this is usually an intermediate stage leading to a comparison with measured data and hence to a confirmation or otherwise of the interpretation.

The author is with the Swinden Laboratories of BSC

548. 73:518. 5

Computers can do two things: they can add or subtract numbers and hence perform a wide range of arithmetical and trigonometrical operations which can be broken down into simple steps, and secondly, they can perform simple logical functions such as deciding whether one number is greater than another or whether an operation results in a positive or negative result. Because they can do these operations with great speed they can be of tremendous help in nearly all aspects of crystallography, which is a branch of science heavily dependent upon computation.

TYPES OF COMPUTER APPLICATIONS

The applications to which computers have been applied can be divided into three main categories:

Reduction of information

Because one cannot appreciate the significance of large amounts of information when in the form of individual numbers it is necessary to reduce them down to a few numbers which uniquely describe the situation. For example, the intensities and angular positions of thousands of single crystal diffraction spots can be measured, but mean little until reduced down to the atomic coordinates representing the crystal structure. The hundreds of intensity readings obtained in stepping across a peak can be reduced to three significant quantities: the angular position of the peak maximum, its width, and its integrated area. The first would be used in lattice parameter calculations, the second in line-broadening studies, and the third in structure determinations or in quantitative phase analysis.

Augmentation of information

It is often necessary to calculate detailed numerical values from a small number of parameters in order to be able to compare these with experimental data. Obvious examples of this are the calculation of all the interplanar spacings or angles from the lattice parameters of a crystal. It may be necessary, having calculated large quantities of numbers to further calculate a single parameter which indicates how near the experimental and calculated values agree; the 'R' factor in crystal structure analysis for example.[1] In most cases where large numbers of individual calculations are recorded, this is only an intermediate step in the work, and a reduction back to a comprehensible size is necessary.

One area where extremely large amounts of information can be understood very quickly is where they are displayed in a graphical form, for instance as a contour map or as stereographic projections. Here the computer can be used to display information graphically either by using a graph-plotter linked to the computer, or by programming the normal print out in such a way that the position of the printed-out digit (which itself represents the z coordinate) is in the correct x and y location.

Control of instrumentation

The most advanced applications of the computer to crystallography are where the computer is linked to a crystallographic instrument[2] and controls the instrument settings in accordance with information received from the previous settings or preset instructions. This application is of great significance in that it frees the operator to concentrate on the crystallographic significance of the results rather than spending most of his time simply collecting data.

AREAS OF INTEREST

There are five main areas where crystallography is used to supply information difficult or impossible to obtain by other techniques. These are:-

  (i) crystal structure determination

  (ii) phase identification

  (iii) quantitative phase analysis

  (iv) crystal lattice studies

  (v) orientation relationships.

There have been computer applications[3] in all these areas, and these will be used to illustrate the way in which the computer is used as previously discussed.

Crystal structure determination

Almost from the time of discovery of X-ray diffraction the technique has been used to solve crystal structures, and more and more complex structures have yielded as time has gone on until at the present time we have the solutions to the structures of the giant molecules such as horse haemoglobin, DNA, etc. Without the aid of a computer, of course, these structures would never have been solved because of the sheer size of the problem. However, recent developments have taken place in combining the computer with the diffractometer to obtain the initial information. When photographic methods were supplemented by diffractometry the problem arose of controlling the position of the counter so that an efficient system could be obtained. With photographic methods there is always a piece of film ready to record a reflection, but with diffractometry the counter has to be positioned correctly as well as the crystal relative to the primary beam. Mechanical linkages mean that most of the time is wasted measuring background away from reflections and it is better to calculate exactly where each reflection will be and position the crystal and counter accordingly to measure the integrated intensity. The use of pulsed stepping motors to drive the crystal and counter is ideal for digital computer links, since the change in position corresponds to integral numbers of pulses. The neutron diffractometer[4] and the apparatus of Mills *et al.*[5] are examples of such systems.

Having obtained the integrated intensities from the diffraction spots which are related to the coefficients of the Fourier transform, it is necessary to reduce this information to a simple set of atomic coordinates which describes the structure uniquely. As already indicated, this can only be done with a computer in complex structures, though many thousands of simpler structures were solved without them.

There are other aspects of structure determination which have been attempted recently by computer. One is the very old problem of solving an unknown powder pattern. Many attempts have been made at this but no satisfactory universal method has emerged. The problem is solvable for cubic, tetragonal, and hexagonal crystals (with Bunn charts for example) but orthorhombic system). Ito's method [7] works in reciprocal space, as does that of de Wolff,[8,9,10] and recent papers by Visser,[11] Goebel and Wilson,[12] Holrocal space, as does that of De Wolff,[8,9,10] and recent papers by Visser,[11] Goebel and Wilson,[12] Holland and Gawthrop,[13] as well as Russian,[14,15]

French,[16] and American[17] workers, have suggested computer routines for applying the methods based on these and statistical considerations.

Phase indentification

The identification of phases by X-ray powder photography or diffractometry is a very important aspect of crystallographic work and it is therefore not surprising to find attempts to put the ASTM index onto cards or magnetic tape[18] which can be sorted by a computer supplied with the list of 'd' spacings, and intensities of the unknown material. However, the general situation is too complex for a good success rate and additional information is necessary to help with the rejection of cards. The store size is also large (~80k), and the search time not negligible either. Chemical information is necessary to exclude wrong cards.

The basic difficulty of course is that the general quality of information in the ASTM index is not good enough for this type of exercise either in the accuracy of its 'd' spacings or in the quoted intensities of the lines. Much of the data was submitted from cylindrical powder camera measurements with visual estimates of intensities. Weak lines are not recorded, and in many cases extra lines appear which belong to another phase (or possibly from β radiation lines being recorded).

When one searches the ASTM index manually one rejects many cards simply from the chemical composition which one knows cannot be relevant to the problem in hand. Furthermore most laboratories have their own library of photographic standards which can be viewed along with the unknown, and either be used to eliminate each compound mentally as it is identified in a mixture, or confirm the identification indicated by the index.

Considerable improvements in the quality of the index are occurring by the systematic work being carried out by NBS and others and by the better data obtainable from Guinier focusing cameras and standard diffractometers compared with the cylindrical powder cameras.

Simple programs to calculate all values of interplanar spacing above a certain limit have been worked out by many workers and used successfully to check standards by being able to account for all the measured lines. Given the lattice parameters and the crystal system, the program calculates the interplanar spacing by varying h, k, and l in steps until a 'd' spacing limit is reached. The values are then tabulated in decreasing order and printed out. Rules for absent reflections can be incorporated to suppress values if the space group is known.

A further refinement of this is to also calculate the intensities if the atomic structure is known. This has proved particularly useful when powder data are not known. For example, there are many forms of silicon carbide which represent the layered structure with different repeat sequences. As it is not possible to control the stacking sequences it is not possible to prepare pure samples of the different forms and therefore one has to fall back on calculation.[19,20]

In electron diffraction, ring patterns can be obtained from powders and identified in the same way as for X-ray patterns. Intensities can be different as the scattering factors for electrons are different from those for X-rays, but can be calculated from the same program. Many electron diffraction patterns, however, arise from single crystal areas and give rise to spot patterns corresponding to slices through the reciprocal lattice. Interzonal angles are thus also of use in identifying the pattern as well as the orientation of the crystal relative to the electron beam. In the case of cubic crystals the angles are independent of lattice parameter,[21] but crystals from less symmetrical systems depend upon the axial ratios and unit cell angles. Each such crystal has its own unique set of angles which can be calculated.[22]

Quantitative phase analysis

To successfully carry out quantitative phase analysis the bounds of the system must be known, i.e. the compounds that could be present and their associated diffraction patterns. The applications vary in complexity from two allotrope systems such as austenite in ferrite or anatase/rutile mixtures to complex systems such as cement clinker. On-line diffractometry is now possible,[23] and therefore process control via diffraction information is possible if crystal structure of the product is the important characteristic.

In systems where lines of the phases are separate it is only necessary to measure the integrated intensities (or peak heights if the peak shape is constant) of the lines, and compare them with those measured on standard mixtures. Where there is extensive overlap of lines it is more difficult, and either the overall profiles have to be separated into different peaks, or sets of simultaneous equations solved to find the contribution of each compound to the peak intensity.

In a typical system, for example, the analysis of cement clinker as attempted by the Cement and Concrete Association, the compounds likely to be present are known and can be prepared in a representative form. The strong lines of many of these phases, e.g. tri- and di-calcium silicates, various calcium aluminates and alumino-ferrites, free lime and spinels occur very close to one another and it is necessary to step scan over an angular range to obtain the peak maxima. These are composed of the sum of the contributions from the individual compounds at each step and hence it is possible to set up a set of simultaneous equations representing the total measured contributions.[24] The solution of these equations gives the individual peak intensities which can be converted to concentrations of crystalline phases with the aid of the standard mixtures information.

The quantitative phase analysis of mixtures emphasises a weakness of diffractometry which can only be overcome by quite sophisticated instrumentation. The diffractometer is very wasteful of time as it is only recording a very narrow band of diffracted information at any instance and therefore to step scan over a large angular range requires many hours of exposure. For example, to scan from 15° to 165° 2θ in 0·02° steps with 30s integration time requires at least 62·5h most of which is spent measuring the background intensity. Hence, a device is required which can distinguish between a peak and the background and speed up the scan rate when on background. Clearly a computer-controlled system is likely to be able to cope best with this situation.

An alternative solution is to record photographically on a Guinier focusing camera which has as good a resolution as a diffractometer (and lower background usually), and microdensitometer the film with an instrument equipped with punch tape output.[25,26,27]

Crystal lattice studies

Information about the crystal lattice, e.g. its size, whether it has imperfections in it or shows solid solution effects, etc., is contained in the line positions and shapes. To determine these it is necessary to scan or step scan across peaks and measure the characteristics of the shapes. Peak positions can be found from a peak-seeking routine and it may be necessary to split a peak into two as in the case of $\alpha_1 \alpha_2$ doublets.[28] In manual operation, lattice parameters can be calculated from the Bragg angles corresponding to each reflection if the symmetry is known, and plotted against one of the usual extrapolation functions to give the parameters at 90° θ. In computer terms a best fit line through the data can be found using standard statistical programs and the value of this at $\theta = 90°$ can be found. These can include weightings if necessary based on Hess's method.[38] Where an internal standard is incorporated the reverse process can be used, i.e. the calculated 'd' spacings for the standard compared with the accepted values to find the amount of correction at each angular position. From this information the unknown material's spacings can be corrected. Such a routine would be used before attempting to index the pattern for example. For very accurate work the centroid of the line should be used rather than the peak maximum.

In measuring line broadening which arises from a combination of instrumental, particle size, and lattice strain broadening, it is necessary to firstly measure line widths, separate any line overlap effects (e.g. $\alpha_1 \alpha_2$ doublets), and remove instrumental broadening.[29] Programs based on Rachinger's[28] method have been used successfully to do this. Programs to calculate the variance of line profiles have also been described[30] as well as line broadening analysis.[31]

Orientation relationships

Much work is done, particularly in electron microscopy, to determine the orientation of one phase relative to another. One phase may be precipitating internally inside the other or growing at the boundary of it. A knowledge of the orientation relationship enables the underlying growth mechanism to be understood more clearly.

To find an orientation relationship it is necessary to index the patterns of both phases and hence find their orientation relative to the electron or X-ray beam direction and hence to each other.[31] The use of graph-plotting routine to draw out stereographic projections[32] in various orientations is very useful in this respect, and Fig. 1 shows such a plot. If the crystal structures of both phases are known it is also possible to compare the atomic arrangements in parallel directions in each phase.[33] A program to draw the atom projections or intersections on any plane has been made[34] to do this. Figure 2 shows an intersection between the 110 plane and the atoms in sigma phase (Fe/Mo).

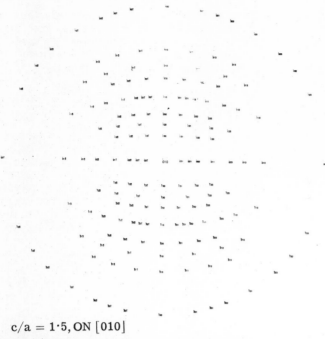

c/a = 1·5, ON [010]

1 Stereographic projection for the tetragonal crystals

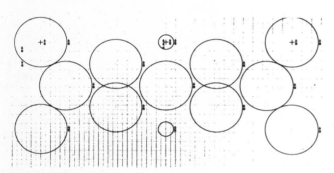

2 110 section of Sigma Phase (Fe/Mo)

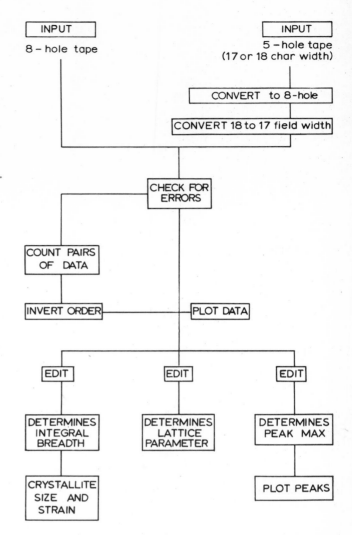

3 Interrelationship of programs (Dr I. Ferguson, UKAEA)

Another obvious example of where graph plotting or visual presentation is of help is in preferred orientation studies.[35,36,37] These involve the collection of considerable quantities of data either by micro-densitometry of sets of exposures taken at varying angles to the specimen plane, or intensity recordings made by a counter which moves in a predetermined way relative to the specimen to cover the range of angles. What is required is a pole density plot on a stereographic projection (or better an equal area projection) and this can be done most effectively by computer, with a graph-plotting attachment.

PRACTICAL CONSIDERATIONS

Although it is relatively easy to design a program to calculate a particular series of operations, the writing of the program is more difficult because of the safeguards needed to be incorporated, and one is advised to obtain a copy of any relevant program already worked out. The situation can be illustrated by the following example. A typical flow sequence worked out by Dr.I. Ferguson[19] is shown in Fig. 3, and apart from having programs to calculate specific operations, it has to take account of these facts:

(i) that data may be introduced in Fortran or Algol in either 5-hole or 8-hole form

(ii) errors may occur in the punching machine, or the tape may be damaged

(iii) if two goniometers on the same X-ray tube are used one has to run backwards and therefore the information on the data tape has to be turned around

(iv) the editing out of selected portions of larger tapes which relate to particular Bragg reflections

(v) varying field widths.

SUMMARY AND CONCLUSIONS

Computers have been applied to many areas of crystallography and basic program problems solved, although the regular application is still confined to a comparatively small number of laboratories. The application of computers to control equipment in crystallography is still not widely applied.

In most cases the computer performs the function of digesting and reducing the amount of information except where visual display is intended. Where large amounts of information are calculated by the computer this is usually an intermediate stage leading to a comparison with measured data and hence to a confirmation or otherwise of the interpretation.

ACKNOWLEDGMENTS

The author wishes to thank Dr F. H. Saniter, OBE, Director of Research, BSC Special Steels Division, for permission to publish this paper. He would also like to thank Mr D. J. Dyson for helpful discussion of program details.

REFERENCES

1 A. D. Booth: *Nature,* 1947, **160**, 196
2 H. C. Freeman *et al.*: Acta Cryst., **A26**, 149
3 D. J. Dyson: *Micron,* to be published
4 A. W. Pryor *et al.*: *J. Appl. Cryst.,* **1**, 272
5 Hilger and Watts Four-Circle Single-Crystal X-Ray Diffractometer, designed by Mills, Edwards, Bowden and Standeven
6 H. Lipson: *Acta Cryst.,* **2**, 43
7 T. Ito: *Nature,* 1949, **164**, 755
8 P. M. de Wolff: *Acta Cryst.,* **10**, 590
9 P. M. de Wolff: *Acta Cryst.,* **11**, 664
10 P. M. de Wolff: *J. Appl. Cryst.,* **1**, 108
11 J. W. Visser: *J. Appl. Cryst.,* **2**, 89
12 J. B. Goebel and A. S. Wilson: USAEC Research and Development Rep., BNWL-22
13 M. J. Holland and J. A. Gawthrop: *J. Appl. Cryst.,* **2**, 81
14 I. A. Barabash *et al*: *J. Appl. Cryst.,* **2**, 10
15 I. A. Barabash *et al*: *J. Appl. Cryst.,* **2**, 14
16 D. J. Taupin: *App. Cryst.,* **1**, 178
17 H. M. Haendler and W. A. Cooney: *Acta Cryst.,* **16**, 1243
18 G. G. Johnson and V. Vand: *Industrial Eng. Chem.,* **59**, 19
19 I. Ferguson: Private Communication
20 D. K. Smith: *Norelco Reporter,* 1968, **15**, 57
21 K. W. Andrews *et al.*: 'Interpretation of Electron Diffraction Patterns', 1967, London, Hilger and Watts
22 D. W. Hogan and D. J. Dyson: Interplanar Angles in the Cubic, Tetragonal and Hexagonal Crystal Systems', 1970, London, Structural Publications
23 G. H. Barton *et al.*: *Trans. IMM,* **78**, 154
24 C. Ballantyne and A. E. Moore: Private Communication from the Cement and Concrete Association
25 'The Autodensidater', Joyce Loebl and Co.
26 W. A. Wooster: *Acta Cryst.,* **23**, 714
27 S. Abrahamsson: *Acta Cryst.,* **A25**, 158
28 W. A. Rachinger: *J. Sci. Inst.,* **55**, 254
29 A. R. Stokes: *Proc. Phys. Soc.,* **61**, 382
30 R. J. Hilleard and J. A. Webster: *J. Appl. Cryst.,* **2**, 193
31 M. S. Rashid and C. J. Altstetter: *J. Appl. Cryst.,* **3**, 120
32 G. K. Stokes *et al.*: *J. Appl. Cryst.,* **1**, 68
33 D. J. Dyson *et al.*: *Acta Met.,* **14**, 867
34 J. C. Minor and D. J. Dyson: *J. Appl. Cryst.,* **3**, 185
35 J. L. Alty: *J. App. Phys.,* **39**, 4189
36 G. R. Love: *Trans. AIME,* **242**, 762
37 A. J. Heckler *et al.*: *Trans. AIME,* **239**, 1241
38 J. B. Hess: *Acta Cryst.,* **4**, 209

DISCUSSION OF THE THIRD SESSION

In the chair: Dr T. Gladman (BSC, Special Steels Division, Rotherham)

Mr J. H. Woodhead (Sheffield University): I was extremely interested in the paper by Professor Nutting and his colleagues on the Scheil type of analysis. I think that most of us who have been concerned with quantitative metallographic measurements have long been aware that this etching problem might be a serious one. Certainly it is known that it has led to an over-estimate of most of the quantities we wanted to measure, but on the whole, we have tended to shrug our sholders and say, 'Let's etch as lightly as possible and hope for the best'.

However, it is very nice to see someone at last has got round to doing something about it, and apparently pretty successfully.

I have only one, perhaps minor, question about the technique, and I wonder if the authors have thought about it. They have made an assumption that you have an etching plane and that this is flat up to the particle. I feel that this is perhaps not too probable and I wonder if they thought at all about the implications of this on the details of the analysis?

In the course of the paper, the authors mention that it would be interesting to investigate the cumulative errors that occur in the computations. I have, in fact, carried out such an investigation and it is hoped that the results will be published in the not-too-distant future. The planar distribution of circles was calculated for a number of idealized spatial distributions of spheres. On the assumption that the observed number of circles in any size group would obey a Poisson distribution, sets of 'observed' data could be calculated by subjecting the true numbers to random deviations obtained by the manipulation of random numbers and the standard deviations relevant to the number of observations in each size group. The resulting 'observed' planar distributions were then converted to spatial distributions by the Scheil method, and from these the variability within size groups was readily obtained.

The results were gratifying, in that provided not less than 1 000 circles were counted, the cumulative errors are nothing like so serious as might be expected. The errors are, of course, greatest for the smallest size group, and if there are relatively few observed particles in this group, it is quite possible to obtain negative numbers from the Scheil analysis.

Dr Hughes: Negative numbers in the smallest group do not really make much difference. They do not make significant contributions to the volume fraction nor, in most cases, do they alter the mean diameter.

The other point about the constancy of etching could affect the distributions in two ways. Firstly, particles which are loosely bonded at the surface may suffer preferential etching and be removed. However, one should get an indication of this on the replicas but we have not observed any such features. Secondly, as regards variable etching rates in different grain orientations, we have prepared plastic replicas which would reveal such an effect as variations in replica thickness from grain to grain. Our observations indicate that although there are small differences they are not serious and the error involved is small compared with the changes introduced by the etching process itself.

Mr Woodhead: It is very satisfying to know that.

Dr J. Barford (CEGB): What shape are the particles?

Dr Hughes: Most of the particles are roughly spherical. There are of course deviations even in apparently isolated particles but the greatest problem is the coalescence of closely spaced particles. In these cases, we usually separate each one into its separate components and measure the equivalent circular sections.

Measurements have been made on the shape parameters of similar Fe_3C particles in ferrite by Heckel and DeGregorio.[*] They deduced that the particles were oblate spheroids with an axial ratio of about 0·9.

Mr Woodhead: I notice in the initial determination of the size distribution, you use a particle size analyser. We have also done this. I wonder if you have ever experienced difficulties when using both size ranges of the instrument to cover a large range of particle sizes, and when it is an advantage to step things up to get the smaller ones more accurately. We find when we do this that we get a discontinuity in the curve where we change from one size range to the other. This appears to be a subjective phenomenon but all our operators get the same sort of effect. Do you find that that?

Dr Hughes: We have not tried that. We always keep to the same scale.

Mr Woodhead: Latterly we too have done this to overcome the problem, but it has been rather disturbing.

---

[*] **R. W. Heckel and R. L. DeGregorio:** *Trans. Met. Soc. AIME*, 1965, **233**, 2001

The effect is not too marked, but it is clearly noticeable.

Mr D. M. Davies (University of Cambridge): Dr Hughes, have you ever considered using a computer to obtain a 'best fit' solution for the size distribution rather than an analytical solution as you have done? This could reduce the uncertainty caused by the negative numbers you obtain with the analytical solution, as it would be possible to specify that only positive numbers were admissable in a 'best fit' solution.

Dr Hughes: No. We are not really worried about these negative numbers since they do not alter the parameters of the distribution by very much.

Mr Davies: I was thinking that if you did obtain a 'best fit' solution you would automatically produce a measure (e.g. x^2) of the errors involved. If the calculation was repeated using the same observed section distribution, but with a range of etch depths in the formulae, then you would obtain the variation of this error value with etch depth. You should observe a minimum in this at the etch depth that was used, e.g. $0.3\ \mu m$ in the sample given in your paper. This would be another indication of the true etch depth.

It is possible to calculate the theoretical distribution using the computer. In Fig. 4, Gilbart *et al.* show theoretical section distributions for an array of equal-sized spheres with three different assumed etch depths. If this calculation were repeated for a number of different sizes corresponding to the mean size in each group, one would obtain a probability function for calculating the size distribution from the observed section distribution. (The details of this method are given by **D. M. Schwartz and B. Ralph:** *Phil. Mag.* **19,** 1061)

I mention this method because it has been used successfully at Cambridge to obtain precipitate size distributions using data from field-ion microscopy. It is difficult to observe many precipitates in any single specimen using this technique, and attempts to use an analytical solution, similar to that used by Gilbart *et al.,* led to solutions with many negative numbers. The non-analytical method, as explained by Schwartz and Ralph, was therefore adopted and has been used successfully on several ferrous-based alloy systems.

The Chairman: It is implicit in this technique that you start fitting at the top end of the size distribution and finish with the misfit at the bottom end, and there is another computer application here, perhaps, in which you can use linear programming. By this technique you obtain the best distribution of sizes which would fit the observations, and it need not necessarily confine the error to the smallest size ranges.

Dr T. B. Vaughan (IMI): I would like to put a question to Mr Johnson, on the last paper on the use of computers in crystallography, with reference to the present state of the art of using computers and graph plotters to produce pole figures.

Some years ago we looked at this. We wrote two programs ourselves, and also tried the ARMCO and the Oakridge programs. We found that the Oakridge was the best, but even making all the economies of time that we could, even by going so far as rewriting the Oakridge program in machine language, we still could not produce stereographic pole figures as cheaply as we could do it by our manual method. This is very much a function of the charging rates for the use of the computer and the graph plotter, and I would guess that in this day and age, one might even question whether one really needs to have pole figures, because there are methods for determining the relative amounts and types of texture without them.

Does Mr Johnson know of laboratories which use graph plotters to produce pole figures, and if so, can they justify the use of this method economically? Is it cheaper than the manual method? May I add a rider to that? If you produce a pole figure by these techniques, do you really get any more useful information out of it than if you simply do it by hand?

Mr Johnson: That is an awkward question. I do not know whether laboratories can justify the use of their computers economically, or not. It is being done and I think that if you have a system where a computer is linked to a diffractometer with a pole figure attachment, then this is the obvious way to do it, but I would agree that this could well be relatively uneconomic. Photographically, you can collect the basic information quickly but if you have a system, with computer tied in, then most of the time is spent simply accumulating counts.

Dr Vaughan: This work we did was three years ago. I wondered if things had altered.

Mr Johnson: I do not think things have altered much.

Mr A. A. Greenfield (Bisra): Our crystallographic people have been doing this using a contour plotting program applied by IBM, but it was a dreadful waste of our computer time. It took an hour or so to plot a pole figure. This was because the contour plotting program works out the value of the function it is plotting and then searches around and finds the next point of the same value and joins up the line. One of my colleagues has written a different type of contour plotting program; it does not matter what the contours are. We are talking about pole figures but it is just a particular contour map. Instead of plotting continuous contours of a function, which could take hours to scan round and join up the points, the program merely evaluates the function across a grid and prints out the values in numerals. It scales the values between 0 and 9 and prints out the numerals which would actually be, as well as scaled in value, also scaled in size. This is done very quickly. It takes only about a minute to do the complete evaluation and printout. You can then take the paper off the graph-plotter and sketch in your contour lines quickly by hand. The choice of method depends on the use of your pole figures. I have asked the crystallography people how they use a pole figure and they say they look at it and it gives them a measure of texture, so I questioned: 'Can you not skip the pole figures; just take the data and shove it into the computer and program it to use whatever method you use in your head when you look at a pole figure to produce a measure of texture.' They say that it is a matter of experience. It is expert judgment that cannot be programmed and they have not been able to produce a technique.

Mr Johnson: Surely you want to take all the information and reduce it down to be able to say finally, for example: 'This particular texture is 10% of {112} ⟨111⟩ with a certain spread, 15% of some other orientation, and 75% of random orientation', and those are the few numbers you really want to be able to ascribe to that material. The fact of drawing a map, etc. is only an intermediate stage.

Mr Greenfield: It is a waste of time.

Mr Johnson: I agree.

Dr W. I. Mitchell (International Nickel Ltd): I have two points. One of them is why have people been wasting their time for years past? It was never thought to be a waste of time in the past, and all of a sudden, when the computer does it, it is a waste of time.

Mr Johnson: This is much more expensive time, I think.

Dr Mitchell: It depends how you cost your assistance. It takes a whole day to do a pole figure, or longer. It does not take more than an hour, surely, even with a fairly sophisticated program, to do a pole figure on a small computer.

Mr Johnson: The drawing of pole figures probably arises from the historic fact that powder photographs were produced where the lines were not uniform circles and this was visually presented on a stereographic projection. This was how it was done originally and it has continued to be done in this way.

Dr Mitchell: I am surprised. I would have thought that pole figures were one thing which could immediately justify the use of computers. Whether you can justify drawing the pole figures is another matter, but you really cannot put that down on the computer.

We have not got any more than a crude pole figure program because we have not sat down to decrudify it. You do not use an assistant; I know that computers are quite expensive but assistants are not cheap.

The Chairman: If you do want a pole figure, then you want to use a computer.

Mr Greenfield: That makes you question why anyone wants a pole figure.

A Delegate: May I ask Dr Vaughan: when he was costing this, was he on-line or off-line? This seems to me potentially a very important question in costing.

Dr Vaughan: We are using a remote computer.

The Delegate: You were on-line the whole time, were-you?

Dr Vaughan: No. What we did was to take the data off the pen chart and simply punch it up in digital form on tape. Obviously, if the thing were feasible economically, we were intending ultimately to install punch tape equipment. From that stage on, putting it through the computer, you can compare the cost of producing a pole figure on a graph plotter, against the time it takes for a man to take the equivalent amount of data off the pen chart and draw it up manually. We were assessing this by producing the equivalent data in punched tape and then putting it into a KDF9 batch type computer to which a graph plotter was attached and we tested different programs. With the charge per unit time on the KDF9 and the charge for the use of the graph plotter, we

found it more expensive than doing it by the manual method by a factor of two even in the best case.

The Delegate: What I was asking was whether you were holding the CPU for the whole time the graph plotter was operating. As it was a KDF9, I presume you were not.

Dr Vaughan: No, a typical pole figure took about 20 minutes calculation time. The information was then stored and put on the plotter subsequently.

Dr D. Goodchild (Uddeholms AB): I am surprised at these estimates of time. I have programmed a KDF9 computer to draw pole figures and it takes about five minutes. We do not actually use the computer to draw in the contours: these are drawn in by hand and we can do this in less than five minutes.

Dr Vaughan: But you have then got to introduce some manual effort.

Dr Goodchild: It takes about five minutes to draw the contours, which is five minutes of manual effort compared with $1\frac{1}{2}-2\frac{1}{2}$ hours required to plot a complete pole figure by hand.

Professor Nutting (University of Leeds): I wonder if Dr Whelan could tell us if he thinks there is any metallurgical information to be got out of the computer-simulated images that could not be obtained from the actual images, or are we to assume that computer simulation is simply a means of confirming that the basic theory of image formation is correct?

Dr Whelan: The object of contrast theory is to try to discover what it is you are looking at in the electron microscope. At present the method used is an indirect one. One makes a guess at the nature of the defect present and then calculates what the image should look like. If this does not agree with what is observed, you modify the guess and try again. There is the possibility of developing the theory in a direct form where the experimental image is used as raw data to deduce directly the form of the defect. A. K. Head has proposed such a form (Australian Journal of Physics). At present the method has not been extensively used.

There is no more metallurgical information in the computer-simulated image than one puts into it.

What other methods are available for extracting useful information from observed electron microscope images? The contrast theory relates the image to fundamental diffraction processes and to the nature of the defect. Without the theory one is just guessing. The state of the theory is good. What is needed now is application of the theory to extract useful information from images. There are many cases where we do not understand what is going on. For example in neutron irradiated metals small specks are observed. Similar defects are observed in proton and ion bombarded speciments. We require to know the nature of such specks, e.g. whether they are point defect clusters and if so whether they are of vacancy, interstitial or impurity type.

The Chairman: An important development in quantitative metallography is the advent of the Quantimet, which is an automated machine for assessing optical microscope images. Recently, we have had a sobering experience while we were studying free-cutting steels. In this exercise, the instrument, automated and working 16 hours a day, produced a vast amount of information. In fact it produced $1\frac{1}{2}$ million digits in the two weeks that it was working, and we were fortunate to foresee that $1\frac{1}{2}$ million digits would require three months for data tape preparation. We were fortunately able to obtain an interface using computer tape, so that we could in fact feed the computer tape out of the Quantimet into the computer, have a print-out of the sheet, check it, and process the results while they were in the computer. Therefore, in terms of automation, you have to be very careful about the sheer volume of data that will be generated.

Dr Whelan: This relates to Dr Duncumb's paper: Is the energy loss along the path based on theory and estimate?

Dr Duncumb: This is using a base relation: energy loss with distance is $^2/_a \times ^1/_e$. It is a continuous energy loss, not discrete.

Dr Whelan: Are the parameters that occur in the electron energy loss law fitted to some experimental measurements of energy losses?

Dr Duncumb: The Monte Carlo calculation is not very sensitive to the precise form of the energy loss law. The mean ionisation potential J which appears in the Bethe relation can be adjusted to fit experimental measurements but does not control the X-ray

absorption very strongly, which is more determined by the scattering conditions assumed. The whole program is not much more than 100 lines long, yet one can get a very good simulation of this very complex process.

The Chairman: In the introductory talk we were trying to define physical metallurgy and we have had a good definition from a physical metallurgist which was subsequently modified to include the effects of commercial characteristics. I think that we are finding applications for computers not only in the industrial field but also in the very fundamental studies. One of the questions raised when considering computer applications is the relevance of the information being sought. This may be a good criterion for a wide range of projects, but it would be dangerous to apply such criteria to the more fundamental studies where the link with profitability may be intangible but very positive in the long term.

Before I close the meeting, I would like to thank the contributors of the papers who have done a really excellent job and also the contributors to the discussion. I think we have had an informal and informative discussion here in Leeds. I would also like to thank members of the ISI staff for their arrangements for the meeting, Professor Nutting for acting as host to this meeting, and also the Physical Metallurgical Committee. I would like you to join me in showing your appreciation in the usual way.

The conference terminated

INDEX OF AUTHORS

*discussion pages

INDEX OF SUBJECTS

---

*discussion pages

---

*discussion pages